# Principles and Analysis of AlGaAs/GaAs Heterojunction Bipolar Transistors

# The Artech House Solid State Technology and Devices Library

*Advanced Semiconductor Device Physics and Modeling*, Juin J. Liou

*Amorphous and Microcrystalline Semiconductor Devices, Volume I: Optoelectronic Devices; Volume II: Materials and Device Physics*, Jerzy Kanicki, editor

*Electrical and Magnetic Properties of Materials*, Philippe Robert

*Computational Modeling in Semiconductor Processing*, M. Meyyappan, editor

*High-Speed Digital IC Technologies*, Marc Rocchi, editor

*InP HBTs: Growth, Processing, and Applications*, B. Jalali and S. J. Pearton, editors

*Indium Phosphide and Related Materials: Processing, Technology, and Devices*, Avishay Katz, editor

*Introduction to Semiconductor Device Yield Modeling*, Albert V. Ferris-Prabhu

*Materials Handbook for Hybrid Microelectronics*, J. A. King, editor

*Microelectronic Reliability, Volume I: Reliability, Test, and Diagnostics*, Edward B. Hakim, editor

*Microelectronic Reliability, Volume II: Integrity Assessment and Assurance*, Emiliano Pollino, editor

*Modern GaAs Processing Techniques*, Ralph Williams

*Numerical Simulation of Submicron Semiconductor Devices*, Kazutaka Tomizawa

*Principles and Analysis of AlGaAs/GaAs Heterojunction Bipolar Transistors*, Juin J. Liou

*Semiconductor Quantum Wells and Superlattices for Long-Wavelength Infrared Detectors*, M. O. Manasreh, editor

*Systematic Analysis of Bipolar and MOS Transistors*, Ugur Çiligiroglu

*Vacuum Mechatronics*, Gerado Beni, Susan Hackwood et al.

*VLSI Metallization: Physics and Technologies*, Krishna Shenai, editor

For further information on these and other Artech House titles, contact:

Artech House
685 Canton Street
Norwood, MA 02062
617-769-9750
Fax: 617-769-6334
Telex: 951-659
e-mail: artech@world.std.com

Artech House
Portland House, Stag Place
London SW1E 5XA England
+44 (0) 71-973-8077
Fax: +44 (0) 71-630-0166
Telex: 951-659
e-mail: bookco@artech.demon.co.uk

# Principles and Analysis of AlGaAs/GaAs Heterojunction Bipolar Transistors

Juin J. Liou

Artech House
Boston • London

**Library of Congress Cataloging-in-Publication Data**
Liou, Juin J.
  Principles and analysis of AlGaAs/GaAs heterojunction bipolar transistors/Juin J. Liou
    p.    cm.
  Includes bibliographical references and index.
  ISBN 0-89006-587-X (alk. paper)
  1. Bipolar transistors. 2. Gallium arsenide semiconductors. 3. Semiconductors—Junctions.
  I. Title.
TK7871.96.B55L56  1996
621.3815'282—dc20                                                    95-53772
                                                                       CIP


**British Library Cataloguing in Publication Data**
Liou, Juin J.
  Principles and analysis of AlGaAs/GaAs heterojunction bipolar transistors
  1. Bipolar transistors
  I. Title
  621.3'815282

  ISBN 0-89006-587-X


Cover design by Christine Koch.


© 1996 ARTECH HOUSE, INC.
685 Canton Street
Norwood, MA 02062


All rights reserved. Printed and bound in the United States of America. No part of this book may be reproduced or utilized in any form or by any means, electronic or mechanical, including photocopying, recording, or by any information storage and retrieval system, without permission in writing from the publisher.
All terms mentioned in this book that are known to be trademarks or service marks have been appropriately capitalized. Artech House cannot attest to the accuracy of this information. Use of a term in this book should not be regarded as affecting the validity of any trademark or service mark.


International Standard Book Number: 0-89006-587-X
Library of Congress Catalog Card Number: 95-53772

10 9 8 7 6 5 4 3 2 1

# Contents

Preface ... ix

Acknowledgments ... xi

Chapter 1 Introduction to the AlGaAs/GaAs Heterojunction Bipolar Transistor (HBT) ... 1
  1.1 Basic Structure and Concept ... 2
  1.2 Heterojunction Properties and Band Discontinuities ... 3
  1.3 Quasi-Fermi Energies in the Heterojunction ... 6
  1.4 HBT Fabrication Technology ... 7
    1.4.1 Non-Self-Aligned HBT ... 8
    1.4.2 Self-Aligned HBT ... 10
    1.4.3 Planar HBT ... 10
    1.4.4 Emmitter-Up Versus Collector-Up Structure ... 11
  1.5 InP- and Si-Based HBTs ... 12
  References ... 15

Chapter 2 Abrupt Heterojunction Bipolar Transistor ... 17
  2.1 Collector Current of Abrupt HBTs ... 17
    2.1.1 Drift-Diffusion Model ... 17
    2.1.2 Thermionic-Field-Diffusion Model ... 18
  2.2 Base Current of Abrupt HBTs ... 21
    2.2.1 Surface Recombination Current ($I_{SR}$) ... 22
    2.2.2 Space-Charge-Region Recombination Current ($I_{SCR}$) ... 25
  2.3 Cutoff Frequency of HBTs ... 29
    2.3.1 Constant Velocity Profile ... 34
    2.3.2 Step-Like Velocity Profile ... 34
    2.3.3 Piecewise-Linear Velocity Profile ... 36
  2.4 Avalanche-Multiplication Characteristics of HBTs ... 37

|  |  |  |
|---|---|---|
| | 2.4.1 Avalanche Collector Current Behavior | 37 |
| | 2.4.2 Reverse Base Current Behavior | 43 |
| 2.5 | Scattering Parameters and Microwave Figures of Merit | 48 |
| References | | 54 |

**Chapter 3  HBTs With Enhanced Structures** — 57
- 3.1 Base Grading — 57
  - 3.1.1 Effect of Base Grading on Current Gain — 58
  - 3.1.2 Effect of Base Grading on Early Voltage — 63
- 3.2 HBTs With a Setback Layer — 64
  - 3.2.1 Collector Current — 67
  - 3.2.2 Base Current — 69
- 3.3 HBTs With a Graded Layer — 72
  - 3.3.1 Collector Current — 79
  - 3.3.2 Base Current — 80
- 3.4 Combined Effect of Setback and Graded Layers — 85
- 3.5 Passivation Emitter Ledge — 92
- 3.6 Proton-Implanted Collector — 92
- References — 97

**Chapter 4  Thermal Effect in an AlGaAs/GaAs HBT** — 99
- 4.1 Self-Heating Effect in Single-Emitter Finger HBTs — 100
  - 4.1.1 Equilibrium Free-Carrier Concentration at High Temperatures — 100
  - 4.1.2 Lattice Temperature and Collector Current Density — 101
- 4.2 Self-Heating and Thermal-Coupling Effects in Multiemitter Finger HBTs — 106
- 4.3 Thermal-Avalanche Interacting Behavior — 117
- 4.4 HBTs Operating Between 300K and 500K — 123
  - 4.4.1 Small-Area HBT — 123
  - 4.4.2 Large-Area HBT — 126
- References — 130

**Chapter 5  Base and Collector Leakage Currents of HBTs** — 133
- 5.1 Leakage Currents at the Emitter-Base Periphery — 135
- 5.2 Leakage Currents at the Base-Collector Periphery — 136
- 5.3 Leakage Currents at the Collector-Subcollector Periphery — 137
- 5.4 Leakage Currents at the Subcollector-Substrate Interface — 137
- 5.5 Base and Collector Currents Including Both Normal and Leakage Components — 138
- References — 148

| | | |
|---|---|---|
| Chapter 6 Noise Characteristics of HBTs | | 149 |
| 6.1 | Overview | 149 |
| 6.2 | High-Frequency Noise Characteristics of HBTs | 152 |
| | 6.2.1 Noise Model | 152 |
| | 6.2.2 Thermal Effect on Noise Behavior | 154 |
| | 6.2.3 Effect of Emitter Area on Noise Behavior | 161 |
| References | | 163 |
| | | |
| Chapter 7 Numerical Simulation of HBTs | | 165 |
| 7.1 | Overview of the MEDICI Two-Dimensional Device Simulator | 165 |
| 7.2 | Effects of Graded Layer, Setback Layer, and Self-Heating | 167 |
| 7.3 | Effects of Different Base and Collector Structures | 178 |
| References | | 195 |
| | | |
| Chapter 8 Reliability Issues of AlGaAs/GaAs HBTs | | 197 |
| 8.1 | Pre- and Post-Burn-In Base and Collector Currents of HBTs | 197 |
| | 8.1.1 Theory | 198 |
| | 8.1.2 Illustrations and Discussions | 199 |
| | 8.1.3 Base Current of HBT Subjected to Long Burn-In Test | 203 |
| 8.2 | Modeling the HBT Current Gain Long-Term Instability | 208 |
| | 8.2.1 Model Development | 209 |
| | 8.2.2 Results and Discussions | 214 |
| References | | 216 |
| | | |
| About the Author | | 219 |
| | | |
| Index | | 221 |

# *Preface*

The role of the silicon bipolar junction transistor (BJT) in integrated circuits has been gradually reduced since the introduction of the metal-oxide-semiconductor field-effect transistor (MOSFET) in the early 1980s. Compared to the MOSFET, the BJT is less compact and consumes more power but possesses a larger transconductance and is thus superior in high-speed applications. In recent years, due to the successful development of molecular beam epitaxy (MBE) and metal-organic chemical vapor deposition (MOCVD), this speed advantage has been further enhanced by the incorporation of a heterostructure into the BJT, resulting in a prominent device called the heterojunction bipolar transistor (HBT). In a typical III-V compound HBT, the base and collector are fabricated with GaAs, and the emitter is made of a material (i.e., AlGaAs) that has a wider energy bandgap than GaAs and a lattice constant and thermal coefficient very similar to that of GaAs. The heterojunction between the emitter and the base allows the base to be doped heavily while maintaining a reasonable current gain. The high base doping concentration thus leads to a small base resistance and a thin base layer, both of which then give rise to a very high cutoff frequency for the HBT.

Many research papers focusing on various aspects of AlGaAs/GaAs HBT have appeared in the literature, but a comprehensive coverage for this increasingly important and popular semiconductor device is not yet available. This book aims to compile important and up-to-date information and to present a unified and concise overview of the physics, analysis, and modeling of AlGaAs/GaAs HBT. The analytical approach will be emphasized, but results obtained from a two-dimensional device simulator are also included. This book is intended to serve as a reference for researchers and engineers working on HBT design, optimization, and characterization. The materials are also highly useful as a supplement for graduate-level semiconductor device courses involving HBT physics and modeling.

Eight chapters are incorporated in the book to provide systematic and up-to-date information on the principles and analysis of AlGaAs/GaAs HBTs. Chapter 1 introduces the basic structure and concept of the AlGaAs/GaAs HBT. Emphasis is placed on the energy band theory in the heterojunction, such as the quasi-Fermi potentials and band discontinuities. A brief description of the HBT fabrication technologies is also given. Furthermore, two other promising technologies, InP- and Si-based HBTs, are briefly discussed, and their advantages and disadvantages as compared to the AlGaAs/GaAs HBT are addressed. In Chapter 2, an abrupt AlGaAs/GaAs HBT, the simplest HBT structure, is treated. Included in this chapter are the discussions of current transport phenomena and performance of the abrupt HBT, such as the dc currents, cutoff frequency, avalanche characteristics, and scattering parameters. This is followed by the analysis of HBTs with enhanced structures in Chapter 3. Commonly used intrinsic and extrinsic structures to enhance the HBT performance are covered. These include the graded base, setback layer after the heterointerface, graded layer before the heterointerface, passivation emitter ledge, and proton-implanted collector. Chapter 4 deals with an important mechanism in an AlGaAs/GaAs HBT called the thermal effect. Both the self-heating effect, which occurs in a single HBT unit, and the thermal-coupling effect, which takes place in an HBT that has multiple emitter fingers, are studied. In addition, the thermal-avalanche interacting behavior of a HBT is investigated. Current transport in a HBT operating at higher ambient temperatures (between 300K and 500K) is also treated in this chapter. Chapter 5 covers the base and collector leakage currents of the HBT—a topic of importance but often overlooked. The leakage currents are normally the dominant components for the base and collector currents at relatively small voltages. The underlying physics of the base and collector leakage currents are addressed, and an analytical model to describe the leakage currents developed. Furthermore, the relevance of the leakage currents to the HBT long-term performance is discussed. The noise characteristics of HBTs are treated in Chapter 6. An overview on the $1/f$, burst, and shot noise in the HBT over a wide frequency range is given. Since the HBTs are used mostly in microwave applications, the high-frequency noise in the HBT is then analyzed in detail. In Chapter 7, numerical studies for single- and multifinger HBTs are carried out using a two-dimensional device simulator called MEDICI. Novel HBT structures are investigated, and their results are compared with those obtained from the conventional structure. Finally, Chapter 8 covers two topics related to HBT reliability. The first is the pre- and post-burn-in current characteristics of the HBT, and the second deals with developing a model to describe the HBT current gain long-term instability.

The main emphasis of the book is device physics and its mathematical representations, through which the operational characterization of AlGaAs/GaAs HBTs can be understood. Each of the chapters contains useful figures to illustrate the trends of the HBTs predicted by the models and, in some cases, those observed in measurements. Fairly extensive references have also been given at the end of each chapter as an aid to the reader who wishes to conduct an in-depth study of a particular subject.

It is my sincere hope that this book will be useful to engineers and researchers who are dealing with HBT problems or are interested in the topic.

# *Acknowledgments*

I am heavily indebted to an anonymous reviewer who offered insightful suggestions and criticisms. Special thanks are due to Doctors C. I. Huang and L. L. Liou at the Wright Laboratory, Wright-Patterson Air Force Base, for their support on the HBT research. Finally I would like to thank my wife Pei-Li, my son William, and my daughter Monica for their understanding and patience during my preparation and revision of the book manuscript.

J. J. Liou
Electrical and Computer Engineering Department
University of Central Florida, Orlando, Florida

## Chapter 1
## *Introduction to the AlGaAs/GaAs Heterojunction Bipolar Transistor*

Despite the higher cost of material and processing, *heterojunction bipolar transistors* (HBTs) have gained popularity in digital and microwave applications primarily because of their superior speed performance. Due to the wide-bandgap emitter used in HBTs, a much higher base doping concentration (e.g., $10^{19}$ cm$^{-3}$) can be used while still maintaining a reasonable current gain. Such a high base doping concentration thus allows the use of a very thin base layer (e.g., 1000 Å) without having to be concerned about punch-through in the base. As a result, the base resistance and the base transit time are reduced and the Early voltage is increased, which leads to a high switching speed and high cutoff frequency (e.g., >100 GHz).

HBT technology is relatively less mature in comparison to *heterojunction field-effect transistor* (HFET) technology, which has benefited from GaAs FET development, but HBTs offer unique performance advantages over silicon *homojunction bipolar transistors* (BJTs) and HFETs. In addition to the high cutoff frequency, HBTs have high current-handling capability and excellent threshold voltage control; they do not suffer from trapping effects, which are the cause of hysteresis in FETs; and they have a wide dynamic range and high output resistance due to a large Early voltage. It is also important to note that the HBT feature size is not limited by the lithography technology, as would be the case for field-effect devices. For field-effect devices, the gate length, which is controlled by how little the window can be opened by the photolithography, is often the limiting factor for the device performance. For HBTs, however, the thickness of the base layer, which is the primary region, needs to be sufficiently wide to prevent punch-through but can be reproducibly fabricated in the neighborhood of 1000 Å or less by epitaxial growth technology such as *molecular beam epitaxy* (MBE) or *metal-organic chemical vapor deposition* (MOCVD).

The idea of utilizing the wide-bandgap emitter to improve device performance was first proposed by Shockley in 1951 [1] and later developed by Kroemer [2, 3]. But most of the developments on the AlGaAs/GaAs HBT, which has become the standard HBT technology, did not start until the early 1980s when the technology for MOCVD and MBE became practical. The selection of the AlGaAS/GaAS pair (normally, $Al_{0.3}Ga_{0.7}As$/GaAs) is due primarily to the fact that AlGaAs and GaAs have very similar lattice constants and thermal coefficients, thus eliminating common mishaps in fabricating heterostructure devices such as substantial lattice mismatch at the heterointerface and thermal cracking. An ultra high speed AlGaAs/GaAs HBT with a frequency of unity current gain (e.g., cutoff frequency) as high as 105 GHz and a maximum frequency of oscillation as high as 175 GHz has been reported [4].

Only the AlGaAs/GaAs HBT will be treated in the book, but many concepts and modeling approaches presented are also applicable to other types of HBTs, such as InP- and Si-based HBTs [5, 6].

## 1.1 BASIC STRUCTURE AND CONCEPT

Consider the simplest HBT structure (single heterostructure, emitter-up structure) consisting of a wide-bandgap AlGaAs emitter (n-type), a heavily doped GaAs base (p-type), a lightly doped GaAs collector (n-type), and a heavily doped subcollector (n-type). These layers are grown on a semi-insulating GaAs substrate by either MBE or MOCVD. The schematic of such a device is shown in Figure 1.1. The energy band diagram of the HBT under forward-active operation (base-emitter voltage $V_{BE} > 0$ and collector-base voltage $V_{CB} > 0$) is given in Figure 1.2, in which $-X_1$ and $X_2$ are the emitter-base space-charge region edges and $X_3$ and $X_4$ are the base-collector space-charge region edges. The superior performance of the HBT results directly from the valence-band discontinuity $\Delta E_V$ at the heterointerface, which stems from the proper choice of the heterojunction pair with emitter bandgap $E_{GE}$ greater than base bandgap $E_{GB}$. $\Delta E_V$ increases the valence-band barrier height in the emitter-base heterojunction and thus reduces the injection of holes from the base to emitter, which is a dominant component of the base current in the silicon homojunction BJT. This allows one to dope the base heavily while maintaining a large current gain. The HBT can then have a very thin base without being concerned about punchthrough in the base region. Thus the benefit of having $\Delta E_V$ is twofold. First, the high base doping concentration decreases the base series resistance, which reduces the ac, dc, and transient emitter crowding [7, 8]; and second, the very thin base reduces the base transit time and enhances the cutoff frequency [9].

The conduction band discontinuity $\Delta E_C$ (spike), on the other hand, is not as desirable as $\Delta E_V$ because the spike makes it necessary for the free carriers in the heterojunction to transport by means of thermionic and tunneling mechanisms [10] instead of drift-diffusion transport in homojunctions, thus impeding the emitter injection efficiency and decreasing the collector current. This problem can be alleviated to some extent by inserting a thin layer (graded layer) before the heterointerface in which the Al mole fraction is

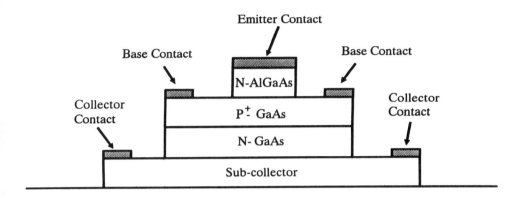

**Figure 1.1** Schematic of the N/p⁺/n AlGaAs/GaAs/GaAs heterojunction bipolar transistor.

graded linearly [10] and/or by inserting a thin, undoped GaAs layer (setback layer or spacer) after the heterointerface [11]. As will be shown in Chapter 3, the junction grading will lower or even remove the spike, thus reducing the importance of thermionic and tunneling mechanisms and making the free-carrier injection from emitter to base more efficient. On the other hand, inserting a setback layer does not alter the spike but rather decreases the barrier potential before the spike. This also makes the thermionic and tunneling mechanisms less prominent and improves the injection efficiency. The spike, however, is sometimes advantageous in that it serves as a launching pad for the electrons. The electrons thus gain considerable kinetic energy before entering the quasi-neutral base. This process results in hot-electron effects and ballistic transport phenomena in HBTs [12, 13].

## 1.2 HETEROJUNCTION PROPERTIES AND BAND DISCONTINUITIES

Let us focus on the emitter-base heterojunction. The conduction band discontinuity $\Delta E_C$ and the valence-band discontinuity $\Delta E_V$ can be fundamentally defined as

$$\Delta E_C = q(\chi_B - \chi_E) \quad \text{and} \quad \Delta E_V = E_{GE} - E_{GB} - q(\chi_B - \chi_E) = \Delta E_G - \Delta E_C \quad (1.1)$$

where subscripts E and B represent the parameters pertinent to the emitter and base, respectively; $\chi$ is the electron affinity, $E_G$ is the energy bandgap; and $\Delta E_G = E_{GE} - E_{GB}$ is the energy bandgap difference. The emitter-base built-in junction voltage $V_{bi}$ is related to the material parameters as [14, 15]

$$V_{bi} = (\chi_B - \chi_E) - (0.5/q)\Delta E_G + V_T \ln(N_E N_B / n_{iE} n_{iB}) + 0.5 V_T \ln(N_{CB} N_{VE} / N_{CE} N_{VB}) \quad (1.2)$$

Here $V_T$ is the thermal voltage, $n_i$ is the intrinsic carrier concentration, $N_E$ and $N_B$ are the emitter and base doping concentrations, and $N_C$ and $N_V$ are the effective density of states in the conduction and valence bands, respectively. Also,

$$V_{B1} + V_{B2} = V_{bi} - V_{BE} \quad (1.3)$$

where $V_{BE}$ is the applied base-emitter voltage and $V_{B1}$ and $V_{B2}$ are the barrier potentials in the emitter and base sides of the junction (Figure 1.2), respectively, and are denoted as

$$V_{B1} = \varepsilon_B N_B (V_{bi} - V_{BE}) / (N_E \varepsilon_E + N_B \varepsilon_B) \quad (1.4)$$

$$V_{B2} = \varepsilon_E N_E (V_{bi} - V_{BE}) / (N_E \varepsilon_E + N_B \varepsilon_B) \quad (1.5)$$

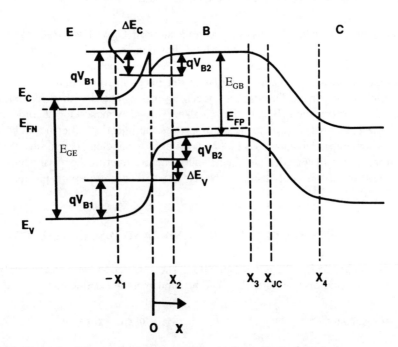

**Figure 1.2** Energy band diagram for the abrupt HBT under forward active operation.

The space-charge region thicknesses on both sides of the heterojunction are

$$X_1 = \{2N_B\varepsilon_E\varepsilon_B(V_{bi} - V_{BE})/[qN_E(N_E\varepsilon_E + N_B\varepsilon_B)]\}^{0.5} \quad (1.6)$$

$$X_2 = \{2N_E\varepsilon_E\varepsilon_B(V_{bi} - V_{BE})/[qN_B(N_E\varepsilon_E + N_B\varepsilon_B)]\}^{0.5} \quad (1.7)$$

Equations (1.4) to (1.7) were derived from the Poisson equation using the conventional depletion approximation and the condition that the potential and the free-carrier flux density are continuous at the heterointerface.

The energy band discontinuities strongly influence the free-carrier transport in the heterojunction, and an accurate determination of such parameters is of primary importance to heterojunction device modeling. Because the exact values of the electron affinity $\chi$ for AlGaAs and GaAs are not known, $\Delta E_C$ and $\Delta E_V$, as defined in (1.1), have to be determined through other means. Unlu and Nussbaum [16] developed a model using the electrostatic approach to predict the conduction and valence-band discontinuities. More recently, an electrochemical approach was proposed by Chang [17] and is considered to be more general and comprehensive compared to its electrostatic counterpart. If $\Delta E_C$ and $\Delta E_V$ are assumed to be doping-concentration independent [18], then both the electrochemical and electrostatic approaches render the same results that $\Delta E_C = 0.21$ eV and $\Delta E_V = 0.19$ eV for an $Al_{0.3}Ga_{0.7}As/GaAs$ heterostructure pair having $\Delta E_G = 0.37$ eV [19]. On the other hand, Frensely and Kroemer [20] and Tersoff [21] suggested that there are heterointerface dipoles associated with states in the bandgap induced by band discontinuities. The magnitude of $\Delta E_C$ and $\Delta E_V$ thus depends on the strength of dipoles, which tends to drive the band lineup toward a condition favorable for zero interface dipole. The drawback of this model is that the number of dipoles is directly proportional to a small difference between two relatively large potentials. As a result, a relative small uncertainty in the potentials can lead to a rather large uncertainty in the dipoles and thus in the band discontinuities. This is especially true for materials involving aluminum.

Many experimental results on $\Delta E_C$ and $\Delta E_V$ have also been reported [22]. In 1974, Dingle et al. [23] determined that $\Delta E_C = 0.85\Delta E_G$ and $\Delta E_V = 0.15\Delta E_G$ based on the quantum-well absorption method. However, the more recent works of Miller et al. [24] and others clearly show that valance-band discontinuity is considerably larger than that proposed in [23]. They suggest the 0.6/0.4 ratio of $\Delta E_G$ for the conduction/valence-band discontinuity (i.e., $\Delta E_C = 0.22$ eV and $\Delta E_V = 0.15$ eV for a $Al_{0.3}Ga_{0.7}As/GaAs$ heterojunction). Figure 1.3 summarizes the results of $\Delta E_V$ measured from the quantum-well absorption method (lower dashed lines) and other methods (upper dashed lines), such as the capacitance-voltage measurement [25] and charge transfer method [26]. Most of the values correspond to 33% to 43% of the bandgap difference between $Al_xGa_{1-x}As$ and GaAs. Throughout the book, the 0.6:0.4 ratio for the conduction/valence-band discontinuity will be used to analyze the HBT.

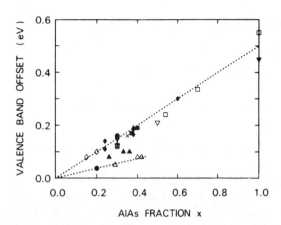

**Figure 1.3**  Valence-band discontinuity $\Delta E_V$ versus Al mole fraction in an $Al_xGa_{1-x}As/GaAs$ heterojunction. The lower and upper dashed lines represent two groups of results based on different measurement methods [26].

## 1.3  QUASI-FERMI ENERGIES IN THE HETEROJUNCTION

In an abrupt n-p heterojunction, the application of a forward bias reduces the barrier to electron flow into the base yet increases the barrier to electron flow in the opposite direction. This leads to a situation where the net electron flow in the heterojunction space-charge region may not be small compared to each of the counterdirected flows. In other words, unlike in a homojunction, the flow of electrons cannot be considered as a small perturbation from equilibrium, and the conventional quasi-equilibrium approximation that the quasi-Fermi level is flat across the space-charge region is not applicable. This fact was first realized by Perlman and Feucht [27] who proposed that the electron quasi-Fermi level be discontinuous at the heterointerface. This is shown in Figure 1.4, where $E_{FNN}$ and $E_{FNP}$ are the electron quasi-Fermi energies in the emitter- and base-side of the space-charge region, respectively. Lunstrom [28] and Pulfrey and Searles [29] later quantified the degree of quasi-Fermi level splitting by using boundary conditions for the electron concentration based on thermionic and tunneling mechanisms at the heterointerface and diffusion in the quasi-neutral base. It should be noted that the quasi-equilibrium approximation still applies to the hole flow and that the hole quasi-Fermi level can be assumed flat across the space-charge region (Figure 1.4).

The separation of the electron and hole quasi-Fermi levels at $x = X_2$ is [29]

$$E_{FNP} - E_{FP} = kT \ln[\Delta n(X_2) + n_0]/n_0 \tag{1.8}$$

where $\Delta n$ is the excess electron concentration and $n_0$ is the equilibrium electron concentration. At the other edge of the space-charge region ($x = -X_1$),

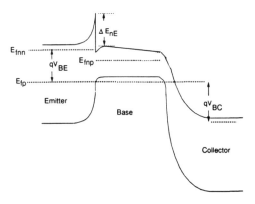

**Figure 1.4** Energy band diagram of the emitter-base heterojunction, with the electron quasi-Fermi level splitting emphasized.

$$E_{\text{FNN}} - E_{\text{FP}} = qV_{\text{BE}} \tag{1.9}$$

Since $E_{\text{FP}}$ is flat across the region, the separation of the electron quasi-Fermi level at $x = X_2$ is thus equal to [29]

$$E_{\text{FNN}} - E_{\text{FNP}} = qV_{\text{BE}} - kT \ln[\Delta n(X_2) + n_0]/n_0 \tag{1.10}$$

Equation (1.10) describes the electron quasi-Fermi level splitting at $x = X_2$, or at the heterointerface ($x = 0$) because $X_2 \approx 0$ due to the high base doping concentration.

Figure 1.5 shows the electron quasi-Fermi level splitting versus $V_{\text{BE}}$ for an abrupt heterojunction, an abrupt heterojunction neglecting tunneling, and a graded heterojunction. For an abrupt heterojunction, the Fermi level splitting increases with $V_{\text{BE}}$ and reaches a value of about 10% of $V_{\text{BE}}$ close to the junction built-in voltage. Also, inclusion of a tunneling mechanism at the heterointerface reduces the Fermi level separation. The results further suggest that the Fermi level splitting is effectively eliminated when the heterojunction is graded.

## 1.4 HBT FABRICATION TECHNOLOGY

Here we review briefly the HBT fabrication methods; a detailed coverage of this subject can be found in [30]. All discussions will be based on the single heterostructure, and three different processing technologies (e.g., non-self-aligned, self-aligned, and planar technologies) will be addressed.

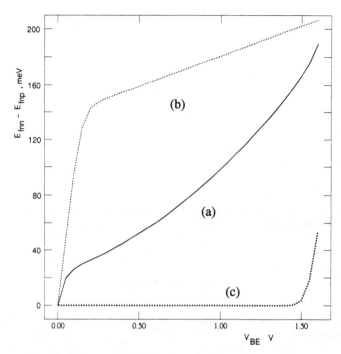

**Figure 1.5** Electron quasi-Fermi level splitting versus the applied voltage for (a) an abrupt heterojunction, (b) an abrupt heterojunction neglecting tunneling effect, and (c) a graded heterojunction (*Source*: [29]. © 1993 IEEE.).

### 1.4.1 Non-Self-Aligned HBT

A non-self-aligned HBT has the following process steps (see Figure 1.6).

1. *Epitaxial Layer Growth:* The subcollector, collector, base, and emitter layers are grown on the semi-insulating GaAs substrate using MOCVD or MBE. For example, if MBE is used, a complex combination of growth temperature, arsenic-to-gallium ratio, base doping, emitter thickness, and possible incorporation of undoped spacer layer needs to be optimized for a given MBE system. A key trade-off is the crystalline quality for p-type dopant diffusion. A higher growth temperature leads to better crystalline quality and low recombination currents but promotes p-type impurity dopant diffusion from base to emitter (out-diffusion effect), especially if Be is incorporated interstitially. A two-step growth temperature technique for the base (lower temperature) and emitter (higher temperature) has been used to optimize the crystalline quality and to reduce the out-diffusion effect. In addition, more recent fabrication has focused on carbon, which is less susceptible to diffusion than Be, for the p-type base dopant [31].

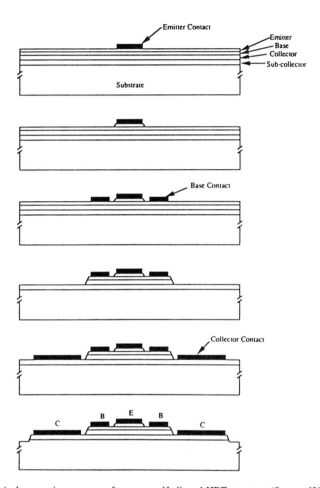

**Figure 1.6** A typical processing sequence for a non-self-aligned HBT structure (*Source*: [30].).

2. *Emitter Contact:* Ohmic contacts are made to the emitter layer by a lift-off technique. Typically, AuGe/Ni/Au, AuGe/Ni/Ti/Au, or Ti/W/Au alloy is used for the emitter ohmic contact.
3. *Base Etch:* Emitter contact is masked with photoresist, and the emitter layer is removed by wet-chemical or dry etching technique to reveal the base layer.
4. *Base Contact:* Base contacts are again made using a lift-off technique. Usually, two base contacts are used, and each base contact is placed as close to the emitter as possible to minimize the base resistance and extrinsic base surface recombination. Although AuBe, AuZn, and AuMg alloys are the most common ohmic contact metals to p-type GaAs, nonalloyed contacts such as Ti/Pt/Au have also been used to prevent spikes extending through the base layer.

5. *Collector Etch:* Collector etching is accomplished by wet or dry etches. After the collector etching, the wafer surface is brought down to the subcollector level.
6. *Collector Contact:* Contacts are made to the subcollector layer by using lift-off techniques and metallization schemes similar to those employed for the emitter contact.
7. *Isolation:* HBTs are isolated from one another by removing the subcollector layer between them. Ion-implantation damage on the subcollector can also be used instead of etching.

### 1.4.2 Self-Aligned HBT

It is often desirable to place the emitter and base contacts as close to each other as possible to reduce the base series resistance and extrinsic base surface recombination. The self-aligned technique is often used to achieve this goal. In such a technique, the base contact is produced in such a way that its separation from the emitter is ensured by a controlled amount because of the emitter geometry. A separation of 0.1 μm to 0.2 μm between the emitter and base contacts is achievable. Figure 1.7 illustrates such a device. For this particular self-aligned structure, the technique uses the emitter metallization as a mask to etch the base layer. The base contact is then aligned to the emitter and the base contact metal is evaporated on the base layer by using the emitter contact as a shadow mask. The base contact is roughly separated from the emitter by the uncut amount.

### 1.4.3 Planar HBT

The HBT fabrication methods just described were for a nonplanar structure, where the entire wafer surface was etched back every time a new layer was accessed in the transistor structure. This results in a tall device structure and creates the need for dielectric cross-overs to connect each device terminal to external circuits.

A device structure can be made more planar by the use of ion implantation

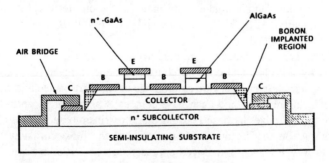

**Figure 1.7.** Cross-section view of a self-aligned HBT (*Source*: [30].).

techniques. Ion implantation is both to form conductive channels and to convert conductive layers into semi-insulating layers. The quasi-planar structure shown in Figure 1.8(a) only utilizes ion implantation for isolation purposes. In the planar structure shown in Figure 1.8(b), device isolation is achieved by implantation as before, but the collector contact is brought to the base layer. Although this structure is more planar, the wafer surface is still at the base layer level. A truly planar structure is not commonly applied to HBTs so as to maintain high emitter injection efficiency. Note that in the planar HBT, the collector contact is no longer self-aligned to the base. The additional collector series resistance that results will lower device performance by dissipating power into this resistance.

### 1.4.4 Emitter-Up Versus Collector-Up Structure

The preceding discussions are limited to the structure where the emitter layer is the topmost layer, which is referred to as the emitter-up structure. An alternative design is the collector-up structure in which the collector is the topmost layer. Figures 1.9(a, b) illustrate the cross-sectional views of typical emitter-up and collector-up HBT structures, respectively.

The major difference between the emitter-up and collector-up structures is the placement of the base layer with respect to the emitter and collector layers. In the case of the emitter-up HBT, the extrinsic base is located directly above the collector, and a

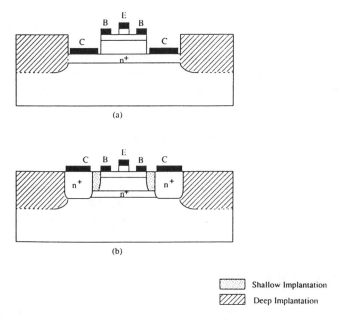

**Figure 1.8** (a) Quasi-planar and (b) fully planar HBT structures (*Source*: [30].).

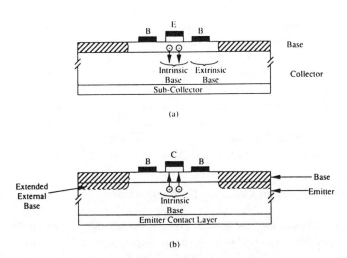

**Figure 1.9** Cross-sectional views of (a) emitter-up and (b) collector-up HBT structures (*Source*: [30].).

majority of electrons entering the base diffuse down toward the collector. As a result, the emitter-up HBT can achieve a very high injection efficiency. In a collector-up HBT, however, the base is located directly above the emitter. Thus, unless special provisions are made, most of the free carriers entering the base are collected at the base contacts. To prevent this from occurring and thus improve the emitter injection efficiency, the external base region boundaries of the collector-up HBT are extended so that a smaller number of free carriers are created at external base boundaries as compared to the intrinsic base region. A commonly used method to achieve this is to convert a portion of the emitter under the external base into a high-resistivity layer by ion-implant damage.

Both structures have their strengths and weaknesses in microwave power amplifier applications. The emitter-up HBT is simpler to fabricate and has a higher emitter injection efficiency. Most microwave-powered HBTs use this structure. The collector-up HBT is preferred for higher frequency operation due to reduced base-collector capacitance. Nonetheless, the collector-up structure is more susceptible to the self-heating effect because the collector layer of such a device sustains the largest portion of the device voltage and most of the heat is generated in this layer. Furthermore, the collector layer is physically isolated from its surroundings (Figure 1.9(b)), which makes the removal of heat generated in the collector-up HBT more difficult. Consequently, the collector-up HBT is not useful in high-power amplifiers but is more suitable for small-signal circuits.

## 1.5 InP- AND SI-BASED HBTs

More recently, two other heterostructure devices also show promising results, that is, InP- and Si-based HBTs. The development of InP-based HBTs focuses on InAlAs/InGaAs and

InP/InGaAs heterojunctions that are lattice-matched to InP substrates. This technology is motivated by the intrinsic advantages of higher free-carrier mobilities, higher heterojunction bandgap discontinuities, lower turn-on voltages, lower surface recombination velocities, and higher substrate thermal conductivity than GaAs-based HBTs [30]. The highest cutoff and maximum oscillation frequencies reported for an InP/InGaAs HBT are 161 GHz and 167 GHz [32], respectively, as shown in Figure 1.10. Furthermore, the higher bandgap discontinuities and reduced surface recombination in an InP-based HBT lead to high dc current gain.

The properties that lead to the high-performance potentials of the InP-based HBTs, nonetheless, also give rise to certain disadvantages, such as lower breakdown voltage, higher leakage currents, and greater difficulties in epitaxial growth and device processing. While AlGaAs/GaAs HBTs benefit from excellent lattice matching (<0.15%) over the full range of Al composition, InP-based HBTs require a restricted compositional combination to achieve lattice matching [33]. Two extensively investigated heterostructure combinations are $In_{0.52}Al_{0.48}As$($E_G$ = 1.44 eV)/$In_{0.53}Ga_{0.47}As$($E_G$ = 0.74 eV) and InP($E_G$ = 1.35 eV)/$In_{0.53}Ga_{0.47}As$. In addition, the choice of the heterostructure system has been limited by the available epitaxial growth technique. InAlAs/InGaAs HBTs are more attractive because the heterostructure can be grown by using the more convenient solid-source MBE.

The Si/SiGe material system, on the other hand, has several properties that make it attractive for HBT applications [6]. Like AlGaAs/GaAs HBTs, Si/SiGe HBTs should have high-speed capability since the base can be heavily doped due to the bandgap difference between silicon and strained SiGe alloys.

Also, the processing of these materials is potentially compatible with existing

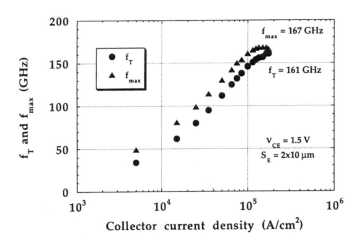

**Figure 1.10** Dependence of cutoff and maximum oscillation frequencies ($f_T$ and $f_{max}$) of an InP/InGaAs HBT with a 2- by 10-μm emitter biased at $V_{CE}$ = 1.5V (*Source*: [32]. © 1995 IEEE.).

silicon technology, although most Si/SiGe HBTs are fabricated using MBE [34]. In addition, the small trapping density at the silicon surface minimizes the surface recombination current and thus ensures a high current gain even at a low collector current density. Figure 1.11(a) shows a typical Si/SiGe HBT structure. A comparison of the base and collector currents measured from a Si/SiGe HBT (with 0.31 Ge mole fraction) and a Si homojunction bipolar transistor (BJT) is given in Figure 1.11(b). The results indicate that the Si/SiGe HBT possesses a higher current gain than the Si BJT. Compared to GaAs- and InP-based HBTs, the Si/SiGe HBT has an inferior cutoff frequency due to the lower free-carrier mobilities in Si.

The major obstacles of the Si/SiGe HBT are the thermal instability and strain relaxation during epitaxial growth. A strained SiGe alloy layer can exist in a metastable state when grown commensurably (with the same lattice constant) on silicon. If the alloy thickness exceeds a certain value, the strain in the SiGe layer will relax and misfit dislocations will form at the heterointerface. Strain relaxation will lower the bandgap difference between Si and SiGe, causing a degradation in emitter injection efficiency and

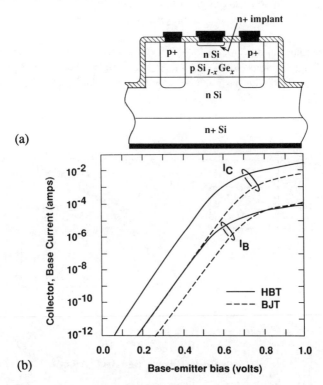

**Figure 1.11** (a) Device structure of an N/p/n Si/SiGe/Si HBT and (b) Gummel plot of a Si/SiGe (with 0.31 Ge mole fraction) and a Si homojunction bipolar transistor (*Source*: [6]. © 1989 IEEE.).

a reduction in current gain. Also, misfit dislocations that accompany strain relaxation can act as generation-recombination centers to increase leakage currents of the HBT.

## References

[1] Shockley, W., U.S. Patent No. 2,569,347, 1951.
[2] Kroemer, H., "Theory of a Wide-Gap Emitter for Transistors," *Proc. IRE,* Vol. 45, 1957, p. 1535.
[3] Kroemer, H., "Heterostructure Bipolar Transistors and Integrated Circuits," *Proc. IEEE,* Vol. 70, 1982, p. 13.
[4] Asbeck, P. M., et al., "Heterojunction Bipolar Transistors for Ultra High Speed Digital and Analog Applications," *IEDM Tech. Digest,* 1988.
[5] Won, T., and H. Morkoc, "Self-Aligned $In_{0.52}Al_{0.48}As/In_{0.53}Ga_{0.47}As$ Heterojunction Bipolar Transistors with Graded Interface on Semi-Insulting InP Grown by MBE," *IEEE Electron Device Lett.,* Vol. 10, 1989, p. 138.
[6] King, C. A., J. L. Hoyt, and J. F. Gibbons, "Bandgap and Transport Properties of $Si_{1-x}Ge_x$ by Analysis of Nearly Ideal $Si/Si_{1-x}Ge_x/Si$ Heterojunction Bipolar Transistors," *IEEE Trans. Electron Devices,* Vol. 36, 1989, pp. 2093–2104.
[7] Houser, J. R., "The Effects of Distributed Base Potential on Emitter-Current Injection Density and Effective Base Resistance for Stripe Transistor Geometries," *IEEE Trans. Electron Devices,* Vol. ED-11, 1964, pp. 238–242.
[8] Yuan, J. S., and J. J. Liou, "Circuit Modeling for Transient Emitter Crowding and Two-Dimensional Current and Charge Distribution Effects," *Solid-St. Electron.,* Vol. 32, August 1989, pp. 623–631.
[9] Liou, J. J., F. A. Lindholm, and B. S. Wu, "Modeling the Cutoff Frequency of Heterojunction Bipolar Transistors Subjected to High Collector-Layer Currents," *J. Appl. Phys.,* Vol. 67, 1990, pp. 7125–7131.
[10] Liou, J. J., "An Improved and Analytical Model for the Current Transport in Graded Heterojunction Bipolar Transistors," *Solid-St. Electron.,* Vol. 38, 1995, p. 946.
[11] Liou, J. J., C. S. Ho, L. L. Liou, and C. I. Huang, "An Analytical Model for Current Transport in AlGaAs/GaAs Abrupt HBTs with a Setback Layer," *Solid-St. Electron.,* Vol. 36, 1993, pp. 819–825.
[12] Maziar, C. M., and M. S. Lundstrom, "On the Estimation of Base Transit Time in AlGaAs/GaAs Bipolar Transistors," *IEEE Electron Device Lett.,* Vol. EDL-8, 1987, pp. 90–92.
[13] Azoff, E. M., "Energy Transport Numerical Simulation of Graded AlGaAs/GaAs Heterojunction Bipolar Transistors," *IEEE Trans. Electron Devices,* Vol. 36, 1989, pp. 609–616.
[14] Chatterjee, A., and A. H. Marshak, "Theory of Abrupt Heterojunctions in Equilibrium," *Solid-St. Electron.,* Vol. 24, 1981, pp. 1111–1115.
[15] Lundstrom, M .S., and R. J. Schuelke, "Modeling Semiconductor Heterojunctions in Equilibrium," *Solid-St. Electron.,* Vol. 25, 1982, pp. 683–691.
[16] Unlu and Nussbaum, "Band Discontinuities as Heterojunction Device Design Parameters," *IEEE Trans. Electron Devices,* Vol. ED-33, 1986, pp. 616–619.
[17] Chang, K.-M., "Band Discontinuities: A Simple Electrochemical Approach," *IEEE Trans. Electron Devices,* Vol. 37, 1990, pp. 883–886.
[18] Liou, J. J., "Correlation Between Electrostatic Approach and Electrochemical Approach for Modeling Band Discontinuities," *Phys. Sta. Sol. (a),* Vol. 125, 1991, p. K21.
[19] Adachi, S., "GaAs, AlAs, and $Al_xGa_{1-x}As$: Material Parameters for Use in Research and Device Applications," *J. Appl. Phys.,* Vol. 58, August 1985, pp. R1–R29.
[20] Frensley, W. R., and H. Kroemer, "Theory of Energy Band Lineup at an Abrupt Semiconductor Heterojunction," *Phys. Rev. B,* Vol. 16, 1977, p. 2642.
[21] Tersoff, J., "Theory of Semiconductor Heterojunctions: The Role of Quantum Dipoles," *Phys. Rev. B,* Vol. 30, 1984, p. 4874.

[22] Wong, W. I., "On the Band Offset of AlGaAs/GaAs and Beyond," *Solid-St. Electron.*, Vol. 29, 1986, p. 133.
[23] Dingle, R., W. Wiegmann, and C. H. Henry, "Quantum States of Confined Carriers in Very Thin $Al_xGa_{1-x}As$-GaAs-$Al_xGa_{1-x}$ As Heterostructures," *Phys. Rev. Lett.*, Vol. 33, 1974, p. 827.
[24] Miller, R. C., D. A. Kleinman, and A. C. Gossard, "Energy-Gap Discontinuities and Effective Masses for GaAs-$Al_xGa_{2-x}$As Quantum Wells," *Phys. Rev. B*, Vol. 29, 1984, p. 7085.
[25] Arnold, D., A. Ketterson, T. Henderson, J. Klem, and H. Morkoc, "Determination of the Valence-Band Discontinuity Between GaAs and (Al,Ga) As by the Use of $P^+$-GaAs-(Al,Ga) As-$P^{1-1}$-GaAs Capacitors," *Appl. Phys. Lett.*, Vol. 45, 1984, p. 1237.
[26] Wang, W. I., E. E. Mendez, and F. Stern, "High Mobility Hole Gas and Valence-Band Offset in Modulation-Doped P-AlGaAs/GaAs Heterojunctions," *Appl. Phys. Lett.*, Vol. 45, 1984, p. 639.
[27] Perlman, S. S., and D. L. Feucht, "p-n Heterojunction," *Solid-St. Electron.*, Vol. 7, 1964, p. 911.
[28] Lundstrom, M. S., "Boundary Conditions for p-n Heterojunctions," *Solid-St. Electron.*, Vol. 27, 1984, p. 491.
[29] Pulfrey, D. L., and S. Searles, "Electron Quasi-Fermi Level Splitting at the Base-Emitter Junction of AlGaAs/GaAs HBT's," *IEEE Trans. Electron Devices*, Vol. 40, 1993, p. 1183.
[30] Ali, F., and A. Gupta, eds., *HEMTs and HBTs: Devices, Fabrication, and Circuits*, Norwood, MA: Artech House, 1991.
[31] Malik, R. J., R. N. Nottenburg, E. F. Shubert, J. F. Walker, and R. W. Ryan, "Carbon Doping in Molecular Beam Epitaxy of GaAs from a Heated Graphite Filament," *Appl. Phys. Lett.*, Vol. 54, 1989, p. 39.
[32] Shigematsu, H., T. Iwai, Y. Matsumiya, H. Ohnishi, O. Ueda, and T. Fujii, "Ultrahigh $f_T$ and $f_{max}$ New Self-Alignment InP/InGaAs HBT's With a Highly Be-Doped Base Layer Grown by ALE/MOCVD," *IEEE Electron Device Lett.*, Vol. 16, 1995, p. 55.
[33] Olego, D., et al., "Compositional Dependence of Bandgap Energy and Conduction Band Effective Mass of InGaAlAs Lattice Matched to InP," *Appl. Phys. Lett.*, Vol. 41, 1982, p. 476.
[34] Patton, G. L., S. S. Iyer, S. L. Delage, S. Tiwari, and J. M. C. Stork, "Silicon-Germanium-Base Heterojunction Bipolar Transistors by Molecular Beam Epitaxy," *IEEE Electron Device Lett.*, Vol. 9, 1988, p. 165.

# Chapter 2
# Abrupt Heterojunction Bipolar Transistor

In this chapter, we discuss and analyze the N/p$^+$/n AlGaAs/GaAs HBT with a standard emitter-up structure including Al$_{0.3}$Ga$_{0.7}$As emitter, GaAs base, and GaAs collector layers (abrupt HBT). The doping concentrations in these layers are uniform, and no additional feature is incorporated. HBTs with advanced structures will be treated in Chapter 3.

## 2.1 COLLECTOR CURRENT OF ABRUPT HBTs

Two approaches have been used in modeling the collector current density $J_C$ in the HBT. One approach is based on the conventional drift-diffusion model, and the other uses the thermionic and tunneling concept at the heterointerface and the diffusion concept in the quasi-neutral base (thermionic-field-diffusion model). As will be shown later, the correct mechanisms of charge transport across an abrupt heterointerface are thermionic emission and tunneling [1, 2], whereas the drift-diffusion model provides reasonable accuracy only if the heterojunction is graded.

### 2.1.1 Drift-Diffusion Model

Using the conventional drift-diffusion concept for the heterojunction together with the assumption that the quasi-Fermi levels are flat and continuous across the space-charge region (quasi-equilibrium approximation), the electron concentration $n(X_2)$ can be modeled as

$$n(X_2) = N_E \exp[-(V_{B1} + V_{B2} - \Delta E_C/q)/V_T] = N_E \exp[(-V_{bi} - V_{BE} - \Delta E_C/q)/V_T] \quad (2.1)$$

Here $N_E$ is the emitter doping concentration; $\Delta E_C$ is the conduction band discontinuity (spike); $V_{B1}$ and $V_{B2}$ are the barrier potentials in the emitter and base sides of the junction, respectively; $V_{BE}$ is the applied base-emitter voltage, and $V_T$ is the thermal voltage. As discussed in Chapter 1, the quasi-Fermi level splitting at the heterointerface is minimal if the junction is graded. As a result, the quasi-equilibrium approximation employed in (2.1) and thus the drift-diffusion model are more accurate for graded HBT and questionable for abrupt HBT. Assuming the recombination current is negligible in the very thin quasi-neutral base, the diffusion-only current at $X_2$ is equal to the drift-only current at $X_3$ ($X_2$ and $X_3$ are the edges of emitter-base and base-collector space-charge regions in the quasi-neutral base; see Figure 1.2). Thus

$$J_C = qD_n[n(X_2) - n(X_3)]/W_B = qv_s n(X_3) \qquad (2.2)$$

where $D_n$ is the electron diffusion coefficient, $n$ is the electron concentration, $W_B = X_3 - X_2$ is the quasi-neutral base thickness, and $v_s$ (= $10^7$ cm/sec) is the saturation drift velocity caused by the high electric field in the base-collector junction. Solving $n(X_3)$ from (2.2) and putting it into the $J_C$ equation yields

$$J_C = qD_n n(X_2)/(W_B + D_n/v_s) \qquad (2.3)$$

### 2.1.2 Thermionic-Field-Diffusion Model

Based on the thermionic concept, the electron current density $J_n$ at the heterointerface ($x = 0$) can be described as the difference of the two opposing electron fluxes [3]:

$$J_n(0) = qv_n[n(0^-) - n(0^+)\exp(-\Delta E_C/kT)] \qquad (2.4)$$

where $v_n$ is the electron thermal velocity, which is denoted

$$v_n = (kT/2\pi m_n^*)^{0.5} \qquad (2.5)$$

Electron concentrations at the vicinity of the heterojunction are

$$n(0^-) = n(-X_1)\exp(-V_{B1}/V_T) \approx N_E \exp(-V_{B1}/V_T) \qquad (2.6)$$

$$n(0^+) = n(X_2)\exp(V_{B2}/V_T) \qquad (2.7)$$

To include the current resulting from the electron tunneling through the triangle-like spike, we need to multiply the electron tunneling coefficient $\gamma_n$ to $J_C$ in (2.4). Thus

$$J_n(0) = qv_n[n(0^-) - n(0^+)\exp(-\Delta E_C/kT)]\gamma_n \tag{2.8}$$

$\gamma_n$ can be expressed as [3]

$$\gamma_n = 1 + \exp(V_{B1}/V_T)(1/V_T)\int_{V^*}^{V_{B1}} D(X)\exp(-V/V_T)\,dV \tag{2.9}$$

where $V^* = V_{B1} - \Delta E_C/q$ for $qV_{B1} > \Delta E_C$ and $V^* = 0$ otherwise, $X = V/V_{B1}$, and $D(X)$ is the barrier transparency

$$D(X) = \exp\{-\eta(1-X)^{0.5} - 0.5\eta X \ln X + \eta X \ln[1 + (1-X)^{0.5}]\} \tag{2.10}$$

Here $\eta = V_{B1}/\{(h/4\pi)[N_E/(\varepsilon_E m_n^*)]^{0.5}\}$ and $h$ is the Planck constant.

Using the diffusion-drift concept in the quasi-neutral base and assuming the recombination current is negligible in the very narrow region of $0 < x < X_2$, $J_C$ can be related to $J_n(0)$ as

$$J_C = \alpha_B J_n(X_2) = \alpha_B J_n(0) \approx qD_n n(X_2)/(W_B + D_n/v_s) \tag{2.11}$$

where $\alpha_B$ is the base transport factor, which will be discussed in the next section. Combining (2.6) to (2.8) and (2.11), we obtain

$$n(X_2) = qv_n\gamma_n N_E\exp(-V_{B1}/V_T)/\zeta \tag{2.12}$$

where

$$\zeta = \alpha_B qD_n/(W_B + D_n/v_s) + qv_n\gamma_n\exp[(V_{B2} - \Delta E_C/q)/V_T] \tag{2.13}$$

Putting (2.12) into (2.1), the collector current density can be calculated from this so-called thermionic-field-diffusion model. If the tunneling current is neglected ($\gamma_n = 1$), then the thermionic-field-diffusion model reduces to the thermionic-diffusion model.

Figure 2.1 illustrates the collector currents for a typical abrupt HBT (device make-up given in Table 2.1) calculated from the drift-diffusion model, from the thermionic-diffusion model, from the thermionic-field-diffusion model, and obtained from measurements. The results suggest that the thermionic-diffusion model is the least accurate and the thermionic-field-diffusion model is the most accurate model among the three models for all voltages. The results also indicate the importance of the tunneling current in the abrupt HBT.

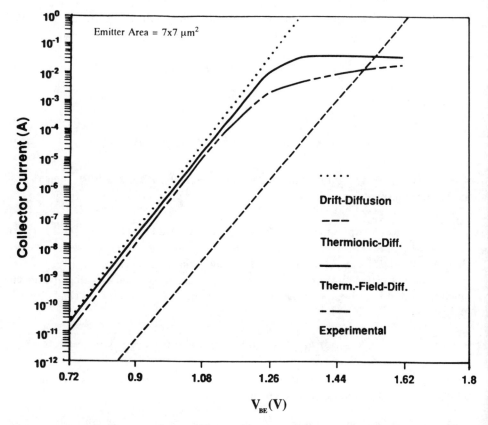

**Figure 2.1** HBT collector current characteristics calculated from the drift-diffusion model, from the thermionic-diffusion model, and from the thermionic-field-diffusion model and obtained from measurement.

**Table 2.1**
HBT Device Structure Used in Calculations

|  | Thickness (Å) | Type | Doping Density (/cm$^3$) | AlAs Fraction in $Al_xGa_{1-x}As$ |
|---|---|---|---|---|
| Emitter | 1700 | n | $5 \times 10^{17}$ | 0.3 |
| Base | 1000 | p$^+$ | $1 \times 10^{19}$ | 0 |
| Collector | 3000 | n | $5 \times 10^{16}$ | 0 |

## 2.2 BASE CURRENT OF ABRUPT HBTs

The trend of the base current of the HBT differs considerably from that of the homojunction BJT. This can sometimes be attributed to the fact that the recombination current in the bulk of the emitter-base space-charge region and the quasi-neutral base is frequently the dominant component of the base current in an HBT, whereas the injection current from the base to the emitter is often attributed most to the base current in a silicon homojunction bipolar transistor. Furthermore, the electron-hole recombination at the surface of a GaAs-related compound is more significant compared to that of silicon. This leads to nonnegligible emitter and base surface recombination currents in AlGaAs/GaAs HBTs due to considerable electron-hole recombination taking place at the surface of emitter and base peripheries [4–9]. Figure 2.2 shows an abrupt N/p$^+$/n AlGaAs/GaAs HBT, where the circles represent the surface states at the extrinsic base surface and the emitter side-walls. The surface recombination currents can become more important than the other base current components when the emitter area is scaled down (e.g., the emitter perimeter-to-junction area ratio is increased) because such currents, unlike the other base current components, are not in direct proportion to the emitter-base junction area. As a result, when the emitter area is scaled down and the surface recombination currents are dominant, the base current remains nearly the same while the collector current decreases, which then reduces the current gain of the HBT.

The base current components of a typical HBT consist of an injection current $I_{RE}$ from the base to the emitter, a recombination current $I_{RB}$ in the quasi-neutral base, a surface recombination current $I_{SR}$ at the surface of extrinsic base and emitter side-walls, and a recombination current $I_{SCR}$ in the bulk of the emitter-base space-charge region. The

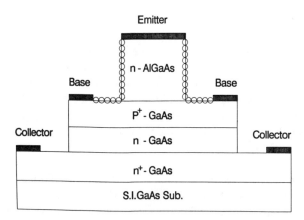

**Figure 2.2** Schematic illustration of an N/p$^+$/n AlGaAs/GaAs HBT structure, where the circles represent the surface states at the extrinsic base surface and at the emitter side-walls.

physics underlying the first two components are very similar to those in the conventional homojunction bipolar transistor. Employing the widely used diffusion-only approximation,

$$I_{RE} = [AqN_BD_p/(X_E - X_1)]\exp[-(V_{B1} + V_{B2} + \Delta E_v/q)/V_T] \quad (2.14)$$

where $A$ is the emitter-base junction area, $N_B$ is the base doping concentration, $D_p$ is the hole diffusion coefficient in the quasi-neutral emitter, $X_E$ is the position of emitter contact, and $\Delta E_V$ is the valence-band discontinuity.

The recombination current in the quasi-neutral base is

$$I_{RB} = AJ_n(X_2)(1 - \alpha_B) \quad (2.15)$$

where $\alpha_B$ is the base transport factor and is expressed as

$$\alpha_B = 1/\cosh[(X_3 - X_2)/(D_n\tau_n)^{0.5}] \quad (2.16)$$

$\tau_n$ is the electron lifetime in the base. Note that $\alpha_B$ approaches unity if the electron-hole recombination in the quasi-neutral base is negligible.

## 2.2.1 Surface Recombination Current ($I_{SR}$)

The surface recombination can take place at the extrinsic base surface and the surface of the emitter side-walls.

*Extrinsic Base Surface Recombination Current ($I_{BS}$)*

Various measurements have shown anomalous current-gain behavior in AlGaAs/GaAs HBTs with exposed extrinsic GaAs base surface. This is due to the fact that substantial electron-hole recombination takes place at the extrinsic base surface. Tiwari et al. [9] investigated such a mechanism using a two-dimensional device simulator. Figure 2.3 shows a perspective plot of volume recombination current density for a typical N/p$^+$/n HBT. In the figure, the base and collector regions extend from $y = 0.352$ to $y = 1.05$ μm. The p-GaAs extrinsic base surface extends from $x = 0$ and $x = 0.5$ μm at $y = 0.352$ μm. It can be seen that the largest amount of recombination occurs at the p-GaAs extrinsic base surface, peaking at the intersection with the base-emitter junction and decreasing toward the base contact. The recombination current at the emitter side-wall, shown vertically to the extrinsic base surface recombination, is somewhat lower than $I_{BS}$ because of the suppression of hole injection by the presence of $\Delta E_V$ at the heterointerface.

The recombination current taking place at the extrinsic base surface is more significant if the base is relatively thick and is not graded [7]. In such a base structure, the minority-carrier flow is not confined to the vertical path, and considerable minority

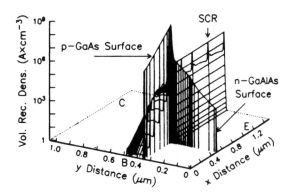

**Figure 2.3** Distribution of recombination current density in the heterostructure bipolar transistor. The locations of the emitter, base, and collector contacts are marked with E, B, and C, respectively [9].

carriers can spread laterally to the surface of the extrinsic base and recombine with holes there via surface trapping states. In a two-dimensional analysis [7], this current was shown to depend on the base thickness, the spacing between the emitter and base electrodes, the size of the emitter electrode, and the density of the surface states. Since $I_{BS}$ depends strongly on processing conditions and is highly two-dimensional, it is difficult to model analytically. Empirically,

$$I_{BS} = I^* \exp(V_{BE}/nV_T) \qquad (2.17)$$

where $I^*$ is the pre-exponential current determined empirically and $n$ is the ideality factor with a value close to unity [10].

*Emitter Side-Wall Surface Recombination Current ($I_{ES}$)*

Let us now focus on the recombination current taking place at the emitter surface (emitter side-walls). To analyze this current, we divide the emitter into $N$ subregions (Figure 2.4) and let the minority-carrier concentration in each subregion be constant (finite-difference approach). Also, the emitter-base space-charge region is assumed to reside only in the emitter side because the base is doped more heavily than the emitter in the HBT.

The emitter region consists of both the quasi-neutral emitter region ($-X_E \leq x \leq -X_1$) and the space-charge region ($-X_1 \leq x \leq 0$) (Figure 2.4). Using the conventional depletion approximation, the space-charge region thickness $X_1$ can be expressed as

$$X_1 = [2\varepsilon_E(V_{bi} - V_{BE})/qN_E]^{0.5} \qquad (2.18)$$

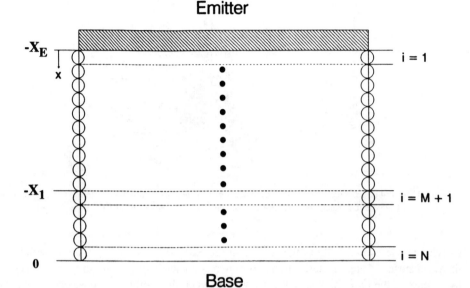

**Figure 2.4** Illustration of the emitter region dividing into $i = 1, \ldots, N$ subregions, where $-X_1$ is the edge of the space-charge region and 0 is the emitter-base heterointerface.

From the above assumptions, the minority-carrier concentration $p^i$ ($i = 1, \ldots, M$, where $M$ is the number of the quasi-neutral emitter subregions) at the surface of each quasi-neutral emitter subregion is [5]

$$p^i = N_B \exp[-(V_B - V_{BE})/V_T](i/M) \quad (2.19)$$

where $V_B$ is the valence-band barrier potential, which is denoted

$$V_B = V_{bi} + \Delta E_v/q \quad (2.20)$$

and $p^i$ ($i = M+1, \ldots, N$, and $N - M$ is the number of the space-charge emitter subregions) in the space-charge emitter subregions is [5]

$$p^i = N_B \exp[-(V_B - V_{BE})/V_T]\exp\{[(V_{bi} - V_{BE})/V_T][(M+1-i)/(M+1-N)]^2\} \quad (2.21)$$

The emitter surface recombination current $I_{ES}$ can be described as

$$I_{ES} = \sum_i (qS^i p^i)(X_E P/N) \quad i = 1, \ldots, N \quad (2.22)$$

where $P$ is the emitter-base junction perimeter and $S^i$ is the surface recombination velocity in each subregion and is denoted

$$S^i = \sum_j [N_{Sj}\sigma_{nj}\sigma_{pj}v_n v_p(n_0^i + p_0^i)]/[\sigma_{nj}v_n(n^i + n_{Tj}) + \sigma_{pj}v_p(p^i + p_{Tj})] \quad (2.23)$$

$j$ is the number of types of surface trap, $N_S$ is the surface trapping density (cm$^{-2}$), $\sigma_n$ and $\sigma_p$ are the electron and hole capture cross sections, $v_n$ and $v_p$ are the electron and hole thermal velocities, $n_0$ and $p_0$ are the equilibrium free-carrier concentrations, and $n_{Tj}$ and $p_{Tj}$ are the electron and hole densities if the equilibrium Fermi level $E_F$ is located at the trap energy $E_T$

$$n_{Tj} = N_{CE}\exp[-(E_C - E_{tj})/kT] \quad (2.24)$$

$$p_{Tj} = N_{VE}\exp[-(E_{tj} - E_V)/kT] \quad (2.25)$$

where $N_{CE}$ and $N_{VE}$ are the effective conduction and valence band density of states and $E_{tj}$ is the energy of traps. Note that $n_0^i \approx N_E$, $p_0^i \approx 0$, and $n^{ia} N_E + p^i$ in the quasi-neutral emitter subregions ($i = 1, \ldots, M$) and that [5]

$$p_0^i = N_B\exp[-V_B/V_T]\exp\{(V_{bi}/V_T)[(M+1-i)/(M+1-N)]^2\} \quad (2.26)$$

$$n_0^i = N_E\exp\{[-(V_{bi} - \Delta E_C/q)/V_T][(M+1-i)/(M+1-N)]^2\} \quad (2.27)$$

$$n^i = N_E\exp\{[-(V_{bi} - \Delta E_C/q - V_{BE})/V_T][(M+1-i)/(M+1-N)]^2\} \quad (2.28)$$

in the space-charge emitter subregions ($i = M + 1, \ldots, N$).

The values of $E_{tj}$ and $N_{Sj}$ at the emitter surface are needed to calculate $S^i$ and thus $I_{ES}$. It has been suggested that there are two dominant traps at the oxided n-type GaAs interface [8]. One type of trap has an energy $E_{t1}$ located at 0.73 eV (relative to the conduction band edge) with a density of $N_{S1} = 5 \times 10^{12}$ cm$^{-2}$ and the other type of trap has energy $E_{t2}$ located at 0.87 eV with a density of $N_{S2} = 10^{11}$ cm$^{-2}$. Assuming these values also apply to AlGaAs, one can calculate $S^i$ and $I_{ES}$ from the above equations with $j = 2$ and $i = N$ ($N$ should be selected based on the convergence and computation efficiencies).

## 2.2.2 Space-Charge-Region Recombination Current ($I_{SCR}$)

Based on the Shockley-Read-Hall (SRH) statistics, the recombination current in the bulk of the space-charge region can be conventionally modeled as

$$I_{SCR} = Aq \int_{-X_1}^{X_2} U_{SRH}\, dx \quad (2.29)$$

where $U_{SRH}$ is the SRH recombination rate. By expanding the integral of (2.29) into the Taylor series and choosing the region contributing to the integral in such a way that the result is exact for the integral $\int_0^\infty \exp(-x^2)\,dx$, Shur [11] derived the following model for $I_{SCR}$:

$$I_{SCR} \approx Aq(\pi/2)^{0.5} V_T n_{iE} \sigma v N_{tB} [qN_E(2V_{bi} - V_{BE})/\varepsilon_E]^{-0.5} \exp(V_{BE}/nV_T) \quad (2.30)$$

where $N_{tB}$ (#/cm$^{-3}$) is the bulk trapping density and $n = 2$ is the ideality factor.

While the above $I_{SCR}$ model is compact and works well for a homojunction, it often fails to describe satisfactorily the space-charge region recombination behavior of a heterojunction. Recently, a comprehensive model [12] was derived rigorously from the integration of the continuity equation over the space-charge region and Shockley-Read-Hall statistics. Their results show the ideality factor $n$ associated with $I_{SCR}$ deviates considerably from that in the conventional model (i.e., $n = 2$). Table 2.2 lists the ideality factors of $I_{SCR}$ for three different base-emitter voltages for abrupt and graded heterojunctions [12]. The bias dependence of the space-charge region recombination current is illustrated in Figure 2.5. Clearly, the ideality factor decreases with $V_{BE}$ for an abrupt heterojunction or a graded heterojunction with a relatively thin graded layer (i.e., 100 Å). The trend becomes less clear if the graded layer is increased to 300 Å.

For illustrations, we consider a typical N/p$^+$/n Al$_{0.3}$Ga$_{0.7}$As/GaAs HBT that has $N_E = 10^{18}$ cm$^{-3}$, $N_B = 2 \times 10^{19}$ cm$^{-3}$, emitter layer thickness of 1000 Å, and base layer thickness of 1000 Å. In the calculations, we neglected the recombination current at the surface of the extrinsic base ($I_{BS} = 0$). Since $I_{ES}$ is a component that does not scale in direct proportion with the emitter junction area, we first illustrate the significance of $I_{ES}$ compared to the rest of the current components for devices with different perimeter-to-area ratios. This is carried out in Figure 2.6 by calculating the ratio $R$ [$R = I_{ES}/(I_{RE} + I_{RB} + I_{SCR})$] for three different emitter areas with a scale down factor of about 5 (Device 1:7 × 7 μm$^2$; Device 2:2 × 5 μm$^2$; and Device 3:1 × 2 μm$^2$). It is shown that the emitter surface recombination current is relatively unimportant at small and very high voltages and is prominent otherwise. Also, the importance of the emitter surface recombination current increases as the ratio of perimeter-to-area increases. As a result, the base current does not scale down in direct proportion with the emitter-base junction area, as evidenced by the

Table 2.2
Ideality Factor of $I_{SCR}$ for Different Voltages and Different Graded-Layer Thickness $W_G$

|  | Abrupt Heterojunction ($W_G = 0$) | Graded Heterojunction ($W_G = 100$ Å) | Graded Heterojunction ($W_G = 300$ Å) |
|---|---|---|---|
| $V_{BE} = 1.0$V | $n = 1.93$ | $n = 1.73$ | $n = 1.33$ |
| $V_{BE} = 1.2$V | $n = 1.80$ | $n = 1.27$ | $n = 1.26$ |
| $V_{BE} = 1.4$V | $n = 1.29$ | $n = 1.24$ | $n = 1.74$ |

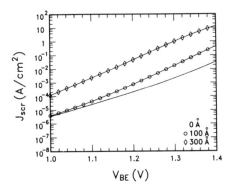

**Figure 2.5** Space-charge region recombination current density versus the base-emitter voltage for three different graded layer thicknesses (*Source*: [12]. © 1992 IEEE.).

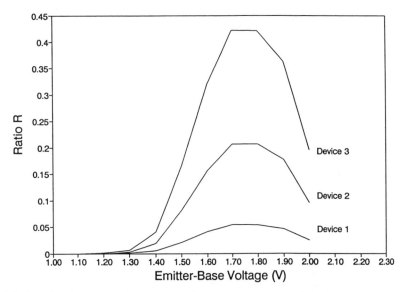

**Figure 2.6** $R$ (ratio of emitter surface recombination current to the rest of the base current components) calculated for Devices 1, 2, and 3.

comparison of the total base currents for the three devices illustrated in Figure 2.7. For example, at $V_{BE} = 1.7$V, the base current is decreased by a factor of 4.68 from Device 1 to Device 2, whereas the base current is decreased by only a factor of 3.8 from Device 2 to Device 3.

A comparison of the base currents calculated from the model without surface current (without $I_{ES}$) at the emitter side-wall and from the model with surface currents for Device 3 is given in Figure 2.8. The results indicate that the current gain of the HBT is degraded

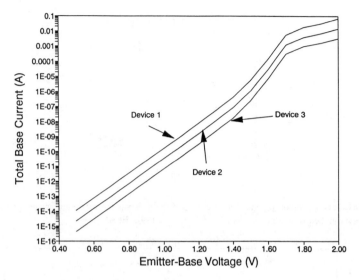

**Figure 2.7** Total base current for Devices 1, 2, and 3.

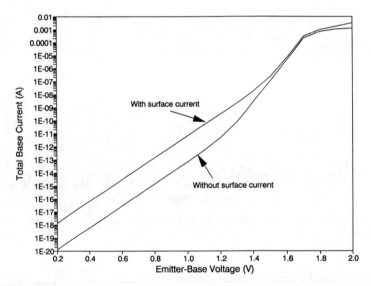

**Figure 2.8** Base currents calculated from the model with and without emitter surface recombination current for Device 3.

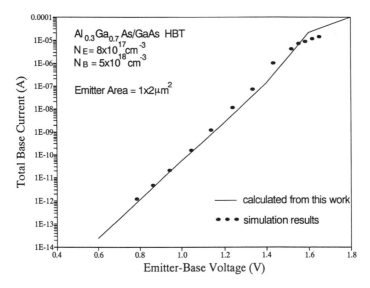

**Figure 2.9** Comparison of the total base current calculated from the present model and obtained from numerical simulation reported in [4].

by surface recombination by a factor of about 100 for $V_{BE} < 1.2$V and about 2 to 5 for $V_{BE}$ above 1.2V. This trend is in agreement with the simulation results reported in [4], which predicted, for a graded-base HBT with a perimeter-to-area ratio of $10^4$ cm$^{-1}$ and a surface recombination velocity of $2 \times 10^6$ cm/s, a decrease in the current gain from 9 to about 0.1 corresponds without and with surface recombination at a collector current density $J_C$ of $10^{-2}$ A/cm$^2$ and likewise from 200 to about 50 at a $J_C$ of $10^5$ A/cm$^2$ (Figures 21(a, b) of [4]). The results in [4] further point out that the current gain of the graded-base device is substantially higher than the device without base grading because the drift field results in fewer electrons injected into the extrinsic base surface. This agrees with our assessment made earlier that extrinsic base surface recombination is reduced if the base is graded.

Simulation results reported in [4] are presented to provide a comparison with the theoretical predications, as shown in Figure 2.9. Good agreement is found.

## 2.3 CUTOFF FREQUENCY OF HBTs

The cutoff frequency $f_T$ is defined as the frequency at which the common-emitter current gain is unity and is an important figure of merit for HBTs used in microwave applications. It is related to four different delay times as

$$f_T = 1/[2\pi(\tau_E + \tau_{BT} + \tau'_{CT} + \tau_C)] \qquad (2.31)$$

where $\tau_E$, $\tau_{BT}$, $\tau'_{CT}$, $\tau_C$ are the emitter charging time, base transit time, collector signal delay time, and collector charging time, respectively. The emitter charge time is

$$\tau_E = r_E(C_{jE} + C_{jC}) \tag{2.32}$$

where $r_E$ is the emitter resistance ($r_E \approx V_T/J_C + r_{EC}$, and $r_{EC}$ is the emitter contact resistance) and $C_{jE}$ and $C_{jC}$ are the emitter-base and base-collector junction capacitances, respectively. Conventionally, $C_{jE}$ and $C_{jC}$ are modeled based on the parallel-plate capacitor concept using the depletion approximation

$$C_{jE} = \varepsilon_E \varepsilon_B/(X_1 \varepsilon_B + X_2 \varepsilon_E) \tag{2.33}$$

$$C_{jC} = \varepsilon_B \varepsilon_C/[(X_{jC} - X_3)\varepsilon_C + (X_4 - X_{jC})\varepsilon_B] \tag{2.34}$$

Note that $\varepsilon_B = \varepsilon_C$ because the HBT under study has a base-collector homojunction.

The diffusion-dominated base transit time, including the base pushout, is

$$\tau_{BT} = (W_B + \Delta W_B)^2/2D_n \tag{2.35}$$

where $W_B = X_3 - X_2$ is the quasi-neutral base thickness and $\Delta W_B$ is the current-induced base pushout [13, 14]. The collector current density $J_0$ for the onset of base pushout can be derived from the Poisson equation using the boundary condition of the electric field at the base-collector metallurgical junction being zero as

$$J_0 = qv_s[N_C + 2\varepsilon_C(V_{bi,bc} = V_{CB})/qX_C^2] \tag{2.36}$$

where $v_s$ is the drift saturation velocity, $N_C$ is the collector doping concentration, $V_{bi,bc}$ is the base-collector junction built-in potential, $V_{CB}$ is the applied collector-base voltage, and $X_C$ is the collector layer thickness. The current-induced base pushout is given by

$$\Delta W_B = X_C\{1 - [(J_0 - qv_sN_C)/(J_C - qv_sN_C)]^{0.5}\} \tag{2.37}$$

for $J_C > J_0$, and $\Delta W_B = 0$ otherwise.

Assuming that the free carriers in the base-collector space-charge layer travel with a drift saturation velocity $v_s$, the collector signal delay time is conventionally modeled as

$$\tau'_{CT} = \tau_{CT}/2 = W_C/2v_s \tag{2.38}$$

where $\tau_{CT} = W_C/v_s$ is called the collector transit time and $W_C$ is the base-collector depletion region thickness.

The collector charging time is

$$\tau_C = r_C C_{jC} \tag{2.39}$$

where $r_C$ is the collector resistance and is denoted by

$$r_C \approx r_{CS} + r_{CC} = X_C/[q(D_n/V_T)N_C] + r_{CC} \tag{2.40}$$

where $r_{CS}$ and $r_{CC}$ are the collector series and contact resistances, respectively.

Let us focus on the collector signal delay time $\tau'_{CT}$, which is often the limiting factor for $f_T$ in a typical AlGaAs/GaAs HBT [15]. The conventional $\tau'_{CT}$ model (see (2.38)) is questionable, however, due to the presence of significant velocity overshoot near the base-collector junction of the HBT [16]. The electric field increases abruptly near the edge of the base-collector depletion layer because the base-collector junction is reverse biased. Because of the very thin base, the electrons entering the base-collector depletion layer are subject to a large local field at a macroscopic point as well as a large gradient of the field. Thus the drift velocity averaged over the ensemble of electrons can exceed those predicted by the conventional drift-diffusion theory, which relies on the assumption that drift velocity is a function of the local field.

Here we developed an improved model for $\tau'_{CT}$ including the velocity overshoot effect. Two different drift-velocity profiles are considered here: (1) the step-like drift velocity profile (Figure 2.10(a)) that has a constant overshoot velocity $V_c$ for the overshoot region ($0 < x < W_o$) and the saturation velocity $v_s$ for the rest of the space-charge layer ($W_o < x < W_C$) [17, 18]; and (2) a more realistic piecewise-linear velocity profile (Figure 2.10(b)) [19], where the overshoot drift velocity $v_o(x)$ is a linearly decreasing function with respect to the position $x$ for $0 \leq x \leq W_o$:

$$v_0(x) = v_p + (v_a - v_p)x/W_o \tag{2.41}$$

where $v_p$ is the peak drift velocity that occurs at $x = 0$ and $v_a$ is the average velocity between $W_o$ and $W_C$. Note that $v_a$ can be larger or smaller than $v_s$ depending on the bias conditions.

Let $x = 0$ be the base-collector metallurgical junction. Following the delay time approach proposed by Ishibashi [18], the induced current $J_{ind}$ through the base-collector junction has the form

$$J_{ind} = J_0 + j(t) \tag{2.42}$$

where $J_0$ is the dc current and $j(t)$ is the ac current. Assuming $j(t)$ has a sinusoidal waveform with an angular frequency $\omega$ yields

$$j(t) = j_0(\sin \omega \tau'_{CT}/\omega \tau'_{CT})\sin \omega(t - \tau'_{CT}) \tag{2.43}$$

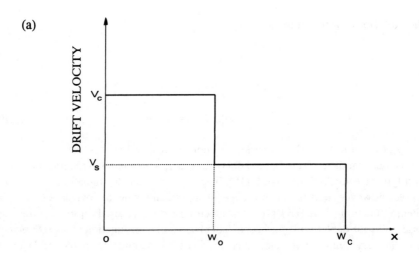

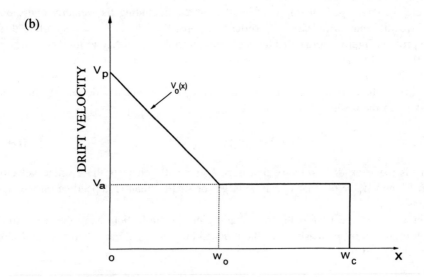

**Figure 2.10** (a) The step-like drift velocity profile consisting of a constant velocity $v_c$ in the overshoot region and the saturation velocity $v_s$ in the saturation region. (b) The piecewise-linear drift velocity profile consisting of a linearly dependent velocity in the overshoot region and an average $v_a$ velocity in the rest of the space-charge region. Depending on the bias conditions, $v_a$ can be smaller or larger than $v_s$.

$J_{ind}$ can also be expressed as

$$J_{ind} = (1/W_C) \int_0^{W_C} [J_0 + j'(t)] \, dx \qquad (2.44)$$

where $j'(t)$ is expressed in terms of the drift velocity in the base-collector space-charge layer as

$$j'(t) = j_0 \sin \omega[t - \int_0^x dy/v_0(y)] \quad \text{for } 0 \leq x \leq W_0 \qquad (2.45)$$

$$j'(t) = j_0 \sin \omega[t - \int_0^{W_0} dx/v_0(x) - (x - W_0)/v_a] \quad \text{for } W_0 \leq x \leq W_C \qquad (2.46)$$

where

$$\int_0^{W_0} dx/v_0(x) = [W_0/(v_a - v_p)] \log_e(v_a/v_p) \qquad (2.47)$$

is the time needed for electrons to travel from $x = 0$ to $x = W_o$.

Incorporating equations (2.42) to (2.47) yields [19]

$$(\sin \omega \tau'_{CT})^2/\omega \tau'_{CT} = (1/W_C)[\int_0^{W_0} \sin \omega C_1 \, dx + \int_{W_0}^{W_C} \sin \omega C_2 \, dx] \qquad (2.48)$$

where

$$C_1 = [W_0/(v_a - v_p)] \log_e\{[(v_a - v_p)x/W_0 + v_p]/v_p\} \qquad (2.49)$$

$$C_2 = [W_0/(v_a - v_p)] \log_e(v_a - v_p) + (x - W_0)/v_a \qquad (2.50)$$

Note that $\tau'_{CT}$ is a function of $\omega$.

For a step-like velocity profile as that used in [17, 18] (Figure 2.10(a)), (2.48) is reduced to

$$(\sin \omega \tau'_{CT})^2/\omega \tau'_{CT} = (1/\omega W_C)\{2v_c \sin(\omega W_0/2v_c) \sin \omega C_3$$
$$+ 2v_s \sin[\omega(W_C - W_0)/2v_s] \sin \omega C_4\} \qquad (2.51)$$

where $v_c$ is the constant velocity in the velocity overshoot region ($0 < x < W_o$) and

$$C_3 = W_0/2v_c \quad \text{and} \quad C_4 = W_0/v_c + (W_C - W_0)/2v_s \qquad (2.52)$$

In the case of low-frequency limitation, $1/\omega$ can be assumed much larger than $W_o/v_c$ and $(W_C - W_o)/v_s$. This leads to

$$\tau'_{CT} \approx \{W_0/v_c + (W_C - W_0)[W_0/v_c + (W_C - W_0)/v_s]/W_C\}/2 \tag{2.53}$$

If a constant velocity profile $[v(x) = v_s]$ is assumed throughout the base-collector depletion region, $W_o = 0$, and (2.51) is further simplified to

$$(\sin \omega\tau'_{CT})^2/\omega\tau'_{CT} = (1/\omega W_C)[2v_s \sin(\omega W_C/2v_s)\sin(\omega W_C/2v_s)] \tag{2.54}$$

Equation (2.54) readily gives $\tau'_{CT} = W_C/2v_s$, which is the conventional model.

Let us turn our attention to (2.48), which is the main result of this section. While obtaining a numerical solution for $\tau'_{CT}$ from (2.48) is theoretically possible, it nonetheless is too tedious to carry out. This is because both integrals in (2.48) involve a sinusoidal function, and accurate numerical integration for this oscillating function requires a very large number of subdivisions between the integral interval. A simplification in the low-frequency limitation can serve to demonstrate the collector signal delay. In the low-frequency case, $\sin(\omega y) \approx \omega y$, where $y$ is a dummy variable, and (2.48) can be written as

$$\tau'_{CT} \approx (1/W_C)[\int_0^{W_0} C_1 \, dx + \int_{W_0}^{W_C} C_2 \, dx] \tag{2.55}$$

We now illustrate and discuss the results of $\tau'_{CT}$ and $\tau_{CT}$ for the constant, step-like, and piecewise-linear velocity profiles as follows.

### 2.3.1 Constant Velocity Profile

In the case of a constant velocity profile, $v(x) = v_s$ throughout the base-collector space-charge region. Thus, according to the foregoing analysis, $\tau_{CT} = W_C/v_s$ and $\tau'_{CT} = W_C/2v_s$.

### 2.3.2 Step-Like Velocity Profile

For the step-like velocity profile as shown in Figure 2.10(a), $\tau_{CT} = W_o/v_c + (W_C - W_o)/v_s$ and $\tau'_{CT}$ can be calculated from (2.53). Figure 2.11(a) plots low-frequency $\tau'_{CT}/\tau_{CT}$ as a function of $W_o/W_C$ for several different $v_c$. $\tau'_{CT}$ equals $0.5\tau_{CT}$ when $W_o = 0$ and $W_o = W_C$, at which the step-like profile reduces to the constant profile. Also shown is that $\tau'_{CT}/\tau_{CT}$ decreases as $v_c$ is increased and that for a particular $v_c$, $\tau'_{CT}/\tau_{CT}$ reaches its minimum value when $W_o/W_C$ approaches about 0.8.

(a)
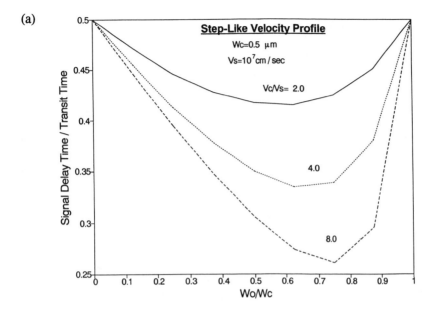

(b)
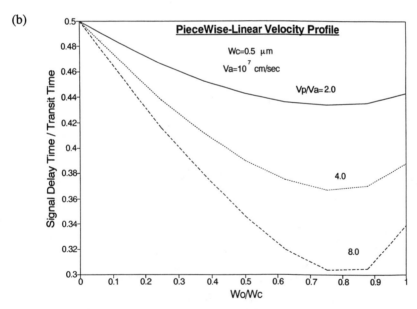

**Figure 2.11** (a) Collector signal delay time versus the width of the overshoot region calculated using the step-like velocity profile. The signal delay time and the width of the overshoot region are normalized by the transit time and by the width of the collector depletion region, respectively. (b) Collector signal delay time versus the width of the overshoot region calculated using the piecewise-linear velocity profile.

### 2.3.3 Piecewise-Linear Velocity Profile

For the piecewise-linear velocity profile (Figure 2.10(b)),

$$\tau_{CT} = W_0 \log_e(v_a/v_p)/(v_a - v_p) + (W_C - W_0)/v_a \tag{2.56}$$

and, in the low-frequency limitation, $\tau'_{CT}$ can be calculated from (2.55). Figure 2.11(b) plots $\tau'_{CT}/\tau_{CT}$ as a function of $W_o/W_C$ for several different $v_p$. $\tau'_{CT}$ differs considerably from $\tau_{CT}$ when $v_p$ is large and when $W_o/W_C$ is increased toward unity. For example, a velocity overshoot that has $v_p = 8v_a$ and $W_o/W_C = 0.5$ will result in $\tau'_{CT}$ being about three times smaller than $\tau_{CT}$, which contrasts with the conventional theory that $\tau'_{CT} = 0.5\tau_{CT}$.

We now consider two collector structures used in [20]. Structure I has a collector doping density of $5 \times 10^{16}$ cm$^{-3}$ and a collector thickness of 0.5 mm. Structure II has a collector doping density of $2 \times 10^{17}$ cm$^{-3}$ and a collector thickness of 0.2 μm. Both devices are Al$_{0.3}$Ga$_{0.7}$As/GaAs HBTs and have identical emitter and base structures. The drift velocity profile needed in modeling $\tau'_{CT}$ (e.g., $v_p$, $W_o$, and $v_a$, which are bias-dependent) can be readily obtained from the simulation results given in Figures 6 and 13 of [20]. Figure 2.12 compares the cutoff frequencies obtained from the particle simulation

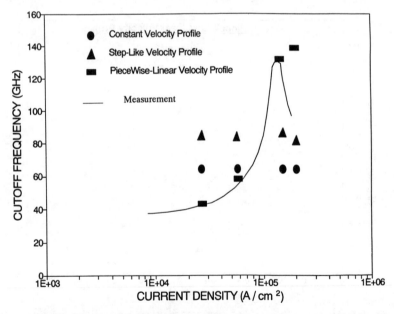

**Figure 2.12** Cutoff frequencies obtained from Monte Carlo simulation (solid line) and calculated from the present approach using the constant, step-like, and piecewise-linear profiles (marks). The calculations are based on the assumption that the collector signal delay time is the dominant component in the total transit time of the HBT. The current densities at which $f_T$ are calculated correspond to the emitter-base applied voltages of 1.5V, 1.53V, 1.55V, and 1.57V from left to right, respectively.

(solid line) [20] and calculated from the model using the constant, step-like, and piecewise-linear velocity profiles (marks) for the Structure I device. Calculations are carried out at the emitter-base applied voltages $V_{BE}$ = 1.5V, 1.53V, 1.55V, and 1.57V, respectively, and are based on the assumption that the total transit time is limited by $\tau'_{CT}$, which leads to $f_T \approx 1/2\pi\tau'_{CT}$. The results indicate that the present approach using the constant and step-like velocity profiles predict constant $f_T$ and nearly constant $f_T$, respectively, for all current densities and that the model using the piecewise-linear velocity profile yields the most favorable results. The falloff of $f_T$ in the range of current density greater than $10^5$ A/cm$^2$ can be attributed to the high-level injection and base pushout.

## 2.4 AVALANCHE-MULTIPLICATION CHARACTERISTICS OF HBTs

Avalanche multiplication is an important mechanism for bipolar transistors because it imposes an upper limit of the collector-emitter voltage for such devices [21–24]. Avalanche multiplication occurs when the reverse-biased base-collector voltage is large. This large voltage results in a large electric field in the base-collector space-charge layer (hereafter called space-charge layer), and free carriers (electrons) that flow through the region are able to gain enough kinetic energy from the field to break covalent bonds in the lattice when they collide with it. This creates two free carriers, that is, a hole and an electron. All these free carriers can then participate in further avalanche collisions, leading to a multiplication of carriers in the space-charge region. Because of the direction of the field, the generated electrons will flow in the same direction as the normally injected electrons. This generated electron current thus gives rise to a sharp increase in the collector current. The generated holes, on the other hand, are drifted into the base region. Since the number of holes that can be injected from the base to emitter is fixed by the emitter-base voltage, these holes are then forced to flow to the base terminal, which reduces the total base current flowing into the terminal. If the avalanche-generated hole current is larger than the normal base current, then a negative base current can result. This is called reverse base current phenomenon [25, 26]. Avalanche multiplication not only depends on the reverse voltage across the base-collector junction, it is also influenced strongly by the charged ion density in the base-collector depletion region, which equals the impurity doping concentration. In addition, avalanche multiplication is affected by the free-carrier charge associated with the collector current passing through the region, which is a function of the base-emitter voltage [27].

### 2.4.1 Avalanche Collector Current Behavior

Consider an N/p$^+$/n HBT under forward-active operation (emitter-base junction forward biased and base-collector junction reverse biased). The collector current $I_C$ including avalanche effects can be expressed as [28]

$$I_C^k = M^{k-1} I_C^{k-1} \qquad k = 1, 2, 3, \ldots, N \tag{2.57}$$

where $M$ is the avalanche multiplication factor and $k$ is the number of the increments of calculations. For example, if an initial value of $I_C$ ($I_C^0$) at an initial applied collector-emitter voltage ($V_{CE}^0$) is known, then $M^0$ at this voltage can be calculated using the model to be developed later. The collector current $I_C^1$ for the next increment of $V_{CE}$ ($V_{CE}^1$) is equal to $M^0 I_C^0$. This current can then be used to calculated $M^1$, which is needed to find IC2. The procedure continues until all $V_{CE}$ increments are calculated.

The avalanche multiplication factor $M$ can be physically expressed as [29]

$$M = 1 + \int_{X_3}^{X_4} \alpha_n \exp[-b_n/\xi(x)]^\beta \, dx \quad (2.58)$$

where $X_3$ and $X_4$ are the left- and right-hand side boundaries of the base-collector space-charge layer (see Figure 1.2), respectively; $\alpha_n$ is the avalanche coefficient; $b_n$ is the critical field; b is the exponential factor; and $\xi(x)$ is the position-dependent absolute electric field in the base-collector space-charge layer. Values for $\alpha_n$, $b_n$, and $\beta$ can be obtained from the literature [30].

The electric field in the space-charge region can be derived from the one-dimensional Poisson's equation

$$d\xi(x)/dx = (1/\varepsilon)[qN_C - J_C/v_s] \quad (2.59)$$

where $\varepsilon$ is the dielectric permittivity, $N_C$ is the collector doping concentration, $J_C = I_C/\text{area}$ is the collector current density, and $v_s$ is the saturation drift velocity. It can be seen from (2.59) that the injected current will tend to neutralize the fixed ion charges in the space-charge region. Thus, if the base-collector voltage is fixed, an increase in the collector current density will increase the thickness of the space-charge region to maintain the charge neutrality in the region. When the current density is sufficiently large, as would be the case in high-level injection, the entire collector can become the space-charge region [13]. As $J_C$ is increased further and is beyond a critical current density $J_0$ at which $\xi(X_3) = 0$, part of the previously depleted collector region adjacent to the quasi-neutral base will become the quasi-neutral region. This in effect can be treated as if the quasi-neutral base has pushed into the collector (base pushout) [31] and $J_0$ is called the onset current density for base pushout.

We now discuss the space-charge layer boundaries and electric field in the region for two possible cases: (1) no base pushout and (2) with base pushout.

*No Base Pushout*

Integrating (2.59) once, we have

$$\xi(x) = (q/\varepsilon)(N_C - J_C/qv_s)(X_4 - x) \quad (2.60)$$

Note that the excess free-carrier (electrons) associated with the collector current flowing through the space-charge layer is accounted for in the model by the term $J_C/qv_s$. Since the base pushout is absent and the base is doped much higher than the collector, $X_3 \approx X_{jC}$, where $X_{jC}$ is the base-collector metallurgical junction. The other boundary of the space-charge layer can be modeled as [13]

$$X_4 = X_{jC} + X_0\{[1 - (J_C/J_2)]/[1 - (J_C/J_1)]\}^{0.5} \quad (2.61)$$

Here $X_0$ is the space-charge layer thickness when $J_C$ is zero, which can be calculated from the conventional depletion model, and $J_1$ and $J_2$ are denoted

$$J_1 = qN_C v_s \quad (2.62)$$

$$J_2 = qN_C \mu_n V_{CB}/W_C \quad (2.63)$$

where $\mu_n$ is the electron mobility, $V_{CB}$ (>0) is the applied collector-base voltage (note that $V_{CE} = V_{CB} + V_{BE}$, where $V_{BE}$ is the applied base-emitter voltage), and $W_C$ is the thickness of the collector. Physically, $J_1$ is $J_C$, at which all electrons move with $v_s$; whereas $J_2$ is the $J_C$ that separates the saturation and active operations. Equation (2.61) indicates that the space-charge layer thickness will increase if $J_C$ is increased, provided $J_1$ is smaller than $J_2$ [13], which is the case for most HBTs biased with a relatively large $V_{CB}$. The electric field given in (2.60) is only applicable for the case when the collector layer is partially depleted ($X_4 < X_{jC} + W_C$).

The collector current density $J_d$ at which the entire collector becomes depleted ($X_4 = X_{jC} + W_C$) can be derived from (2.61) by letting $X_4 = X_{jC} + W_C$ as

$$J_d = [1 - (W_C/X_0)^2]/\{[J_1 - J_2(W_C/X_0)^2]/J_1 J_2\} \quad (2.64)$$

The electric field for $J_C > J_d$ is given by [25]

$$\xi(x) = 0.5(qW_C/\varepsilon)(N_C - J_d/qv_s) + 0.5(qW_C/\varepsilon)(N_C - J_C/qv_s)$$

$$- [(q/\varepsilon)(N_C - J_C/qv_s)(x - X_{jC})] \quad (2.65)$$

The boundaries of the space-charge layer for this case are

$$X_3 = X_{jC} \quad \text{and} \quad X_4 = X_{jC} + W_C \quad (2.66)$$

*With Base Pushout*

The entire collector region can become the space-charge layer if $J_C$ is sufficiently large, and an even higher $J_C$ can induce base pushout. The onset current density $J_0$ for base pushout can be derived by using the boundary condition $\xi(X_3) = 0$, integrating (2.60) once, and letting $J_C = J_0$ at the condition

$$J_0 = qv_s[N_C + 2\varepsilon(V_{CB} + V_{bi,bc})/qW^2c] \qquad (2.67)$$

where $V_{bi,bc}$ is the base-collector junction built-in potential.

The current-induced base widening $\Delta W_B$ is given by [31]

$$\Delta W_B = W_C\{1 - [(J_0 - qv_sN_C)/(J_C - qv_sN_C)]^{0.5}\} \qquad (2.68)$$

for $J_C > J_0$ and by $\Delta W_B = 0$ otherwise. Thus the two boundaries for the space-charge layer are

$$X_3 = X_{jC} + \Delta W_B \quad \text{and} \quad X_4 = X_{jC} + W_C \qquad (2.69)$$

The electric field in the space-charge layer for $J_C > J_0$ is [27]

$$\xi(x) = \{[0.5q(W_C - \Delta W_B)/\varepsilon](N_C - J_C/qv_s)\} + \{[0.5q(W_C^2/\varepsilon)(N_C - J_d/qv_s)]/(W_C - \Delta W_B)\}$$
$$- [(q/\varepsilon)(N_C - J_C/qv_s)(x - X_{jC} - \Delta W_B)] \qquad (2.70)$$

From the above model, one can calculate numerically and successively $M$ and $I_C$ as functions of $V_{CB}$ for a given initial collector current.

To verify the model developed, we use experimental data measured from an Al$_{0.3}$Ga$_{0.7}$As/GaAs/GaAs HBT fabricated in Wright Lab, Wright-Patterson Air Force Base, using a self-aligned process. It has an emitter junction area of $40 \times 40$ μm$^2$, $N_E = 7 \times 10^{17}$ cm$^{-3}$, $W_E = 0.1$ μm, $N_B = 5 \times 10^{19}$ cm$^{-3}$, $W_B = 0.06$ μm, $N_C = 8 \times 10^{15}$ cm$^{-3}$, and $W_C = 1$ μm. The avalanche parameters ($\alpha_n$, $b_n$, and $\beta$) have been investigated extensively in the past, and different results were suggested in the literature [30]. The commonly used values for GaAs are (1) $a_n = 2 \times 10^6$ cm$^{-1}$, $b_n = 2 \times 10^6$ V/cm, $\beta = 1.0$; (2) $\alpha_n = 2.99 \times 10^5$ cm$^{-1}$, $b_n = 6.8 \times 10^5$ V/cm, $\beta = 1.6$; and (3) $\alpha_n = 3.5 \times 10^5$ cm$^{-1}$, $b_n = 6.8 \times 10^5$ V/cm, $\beta = 2.0$. We have tried all three sets of parameters in calculations and found that the first set yields the best agreement with measurements. Figure 2.13 shows the $I_C - V_{CE}$ characteristics calculated from the model and obtained from measurements. Excellent agreement is found between the model predictions and experimental data. The corresponding multiplication factors for the currents are plotted in Figure 2.14. The results suggest that before avalanche breakdown occurs ($V_{CE} < 20$V), $M$ increases monotonically as the voltage is increased and $M$ is smaller if the initial collector current

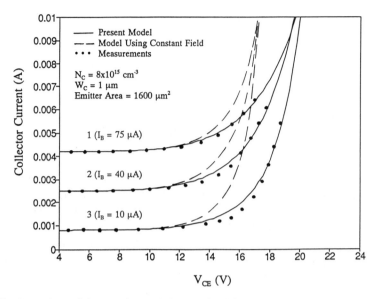

**Figure 2.13** Comparison of the *I–V* characteristics calculated from the present model and obtained from measurements for three different initial collector currents of 0.8 mA, 2.5 mA, and 4.2 mA.

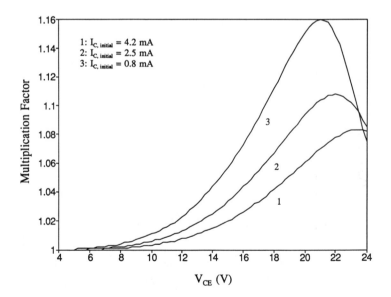

**Figure 2.14** Multiplication factors for the three *I–V* curves given in Figure 2.13.

is higher. The former results because a large voltage gives rise to a large electric field in the space-charge layer and therefore a larger $M$. The latter is due to the fact that a higher current will inject more free carriers (electrons) into the space-charge layer, which compensates the ion charges in the region and thus reduces the electric field. After avalanche breakdown occurs ($V_{CE} > 20$V), however, $M$ decreases as the voltage is increased because the very high current level induces base pushout, which reduces the space-charge layer thickness and therefore the avalanche multiplication factor.

We next examine the effects of grading the collector Al mole fraction on avalanche behavior. Similar to the concept of base grading, the Al mole fraction ($x'$) in the collector can be graded to create a built-in electric field that has a direction opposite that of the electrostatic field in the junction. As a result, the magnitude of the overall field is decreased, which reduces impact ionization and thus avalanche multiplication. Figure 2.15(a) plots $I_C - V_{CE}$ characteristics for three cases: (1) $x' = 0$ (i.e., no grading or the collector material is GaAs); (2) a linear grading of $x' = 0.2$ to 0 over one-half of $W_C$ adjacent to the base-collector junction; and (3) a linear grading of $x' = 0.4$ to 0 over one-half of $W_C$ adjacent to the base-collector junction. No substantial improvement on the avalanche breakdown is found. If the same grading is carried out over a smaller distance (e.g., one-fourth of $W_C$), then it becomes more effective than the previous one, as shown in Figure 2.15(b), but the improvement on the avalanche breakdown is still minimal.

The collector doping concentration, on the other hand, can strongly influence the avalanche behavior, as evidenced in the results shown in Figure 2.16. The corresponding

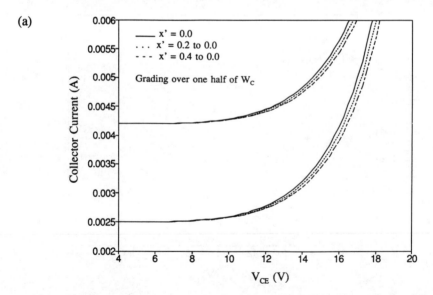

**Figure 2.15** Avalanche current characteristics calculated for two different initial collector currents of 2.5 mA and 4.2 mA and three different Al grading: (a) over one-half of the collector thickness adjacent to the base-collector junction and (b) over one-fourth of the collector thickness adjacent to the base-collector junction.

(b)

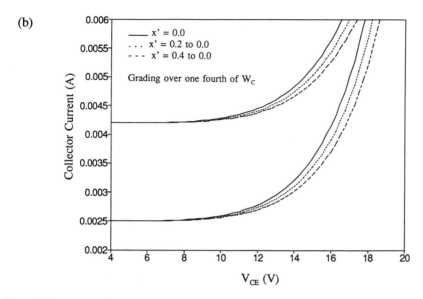

**Figure 2.15** (continued)

multiplication factors are given in Figure 2.17. Again, the fall-off of $M$ beyond avalanche breakdown results from the current-induced base pushout. We have also calculated the $I$–$V$ characteristics for three different collector thicknesses, and the results indicate that reducing $W_C$ will give rise to a smaller breakdown voltage (Figure 2.18). The effect of the different junction areas (a scaling factor of 2.5 is used), which yields different current densities, is also studied. As shown in Figure 2.19, if the junction area is decreased or if the current density is increased, the model predicts that avalanche multiplication takes place at a larger $V_{CE}$, which is confirmed by data measured from a HBT that has the same device make-up as that used in Figure 2.13 but a smaller area of 625 $\mu m^2$, because a higher current density injects a larger number of electrons into the space-charge layer, thus reducing the ion charges and the electric field in the region.

### 2.4.2 Reverse Base Current Behavior

The model developed in Section 2.4.1 can also be used to describe the reverse base current associated with the avalanche multiplication phenomenon. The base current $I_B$ including avalanche effects is given by

$$I_B = I'_B - (M-1)I'_C \tag{2.71}$$

where $I'_B$ and $I'_C$ are the base and collector currents without avalanche effects, respectively. The two currents can be expressed in terms of the applied base-emitter voltage $V_{BE}$ as [32]

$$I'_B = A(qD_pN_B/W_E)\exp[-(V_{bi,be} + \Delta V_C - V_{BE})/V_T] + I^*\exp(V_{BE}/2V_T) \tag{2.72}$$

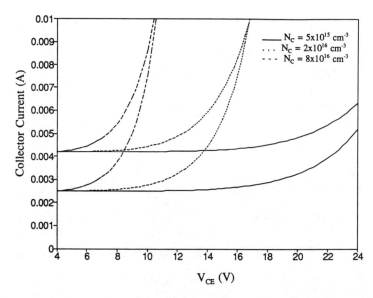

**Figure 2.16** Avalanche characteristics calculated for three different collector doping concentrations.

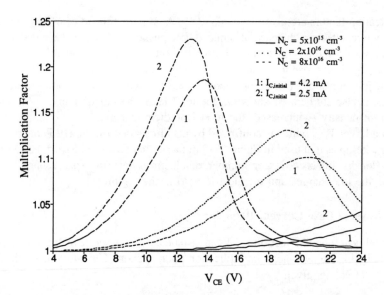

**Figure 2.17** Multiplication factors corresponding to the *I–V* characteristics shown in Figure 2.16.

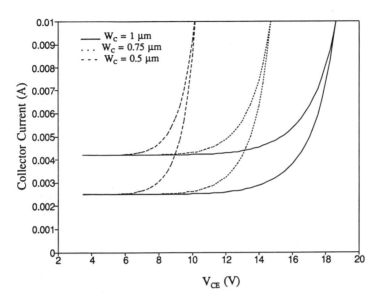

**Figure 2.18** Avalanche characteristics calculated for three different collector thicknesses.

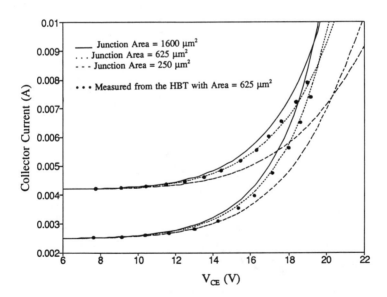

**Figure 2.19** Avalanche characteristics calculated for three different junction areas with a scaling factor of 2.5.

$$I'_C = A(qD_n N_E/W_B)\exp[-(V_{bi,be} - V_{BE})/V_T] \quad (2.73)$$

Here $I^*$ is the pre-exponential factor of the recombination current in the emitter-base heterojunction space-charge layer; $D_p$ and $D_n$ are the hole and electron diffusion coefficients, respectively; $V_{bi,be}$ is the emitter-base junction built-in potential; $W_E$ and $W_B$ are the emitter and base layer thicknesses, respectively; and $\Delta V_C$ is the valance band discontinuity at the heterointerface. Note that $I^*$ is a function of the Shockley-Read-Hall recombination parameters, such as the trapping density and capture cross section, at the heterointerface as well as in the bulk of the emitter-base space-charge layer.

Equation (2.71) indicates that the base current can become negative if the avalanche multiplication factor is large. Figures 2.20(a, b) show the Gummel plots calculated from the present model and obtained from measurements for $V_{CB} = 0.5V$ and $V_{CB} = 16V$, respectively. Good agreement is found between the model predictions and measured dependencies, except at small voltages where an additional dip is observed in measurement (Figure 2.20(b)) that is caused by the leakage currents and the asymmetric emitter-base and base-collector junctions, the effects of which are not accounted for in the model. For the HBT considered, both the calculated and measured dependencies indicate that the reverse base current (negative base current) can occur when $V_{CB}$ is large and $V_{BE}$ is between 1.2V and 1.3V. This results from the relatively large avalanche multiplication factor under such bias conditions (Figure 2.21). The large multiplication factor generates a large number of holes in the base-collector space-charge layer. These holes are then drifted into the base and forced to flow out of the base terminal, resulting in a negative

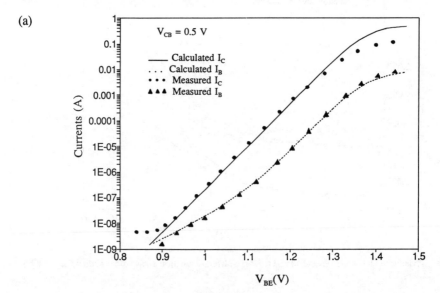

**Figure 2.20** Base and collector currents versus the applied base-emitter voltage for (a) $V_{CB} = 0.5V$ and (b) $V_{CB} = 16.0V$.

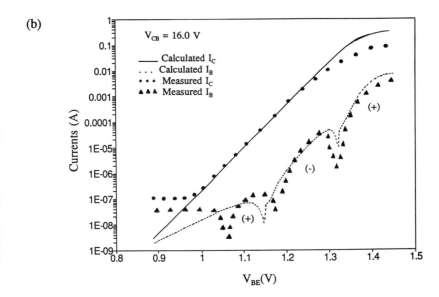

**Figure 2.20** (continued)

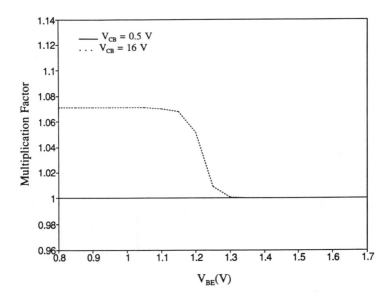

**Figure 2.21** Multiplication factors corresponding to the results shown in Figures 2.20(a, b).

base current [25]. Note that the reduced multiplication factor at large $V_{BE}$ (Figure 2.21) causes the base current to return to the positive value.

## 2.5 SCATTERING PARAMETERS AND MICROWAVE FIGURES OF MERIT

To characterize the HBT microwave performance, scattering parameters (s-parameters) are extensively used because they are easier to measure at high frequencies than other kinds of parameters [33]. Figure 2.22 shows a general two-port network with incident ($a_1$ and $a_2$) and reflected waves ($b_1$ and $b_2$). Using the s-parameter definition

$$\begin{bmatrix} b_1 \\ b_2 \end{bmatrix} = \begin{bmatrix} s_{11} & s_{12} \\ s_{21} & s_{22} \end{bmatrix} \begin{bmatrix} a_1 \\ a_2 \end{bmatrix} \quad (2.74)$$

where the s-parameters are

$$s_{11} = b_1/a_1|_{a2=0} \quad s_{22} = b_2/a_2|_{a1=0} \quad s_{21} = b_2/a_1|_{a2=0} \quad s_{12} = b_1/a_2|_{a1=0} \quad (2.75)$$

Several HBT high-frequency performances are related to the s-parameters. For example, the power gain $G_p$ is a figure of merit for microwave HBT; it is the ratio of power delivered to the load $Z_L$ to power input to the network and can be expressed in terms of the s-parameters as [34]

$$G_p = |s_{12}|^2(1 - \Gamma_L^2)/[(1 - |s_{11}|^2) + \Gamma_L^2(|s_{22}|^2 - D^2) - 2\,\text{Re}(\Gamma_L N)] \quad (2.76)$$

$$\Gamma_L = (Z_L - Z_g)/(Z_L + Z_g) \quad D = s_{11}s_{22} - s_{12}s_{21} \quad N = s_{22} - Ds_{11}^* \quad (2.77)$$

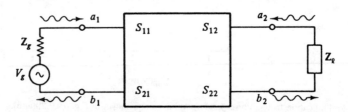

**Figure 2.22** S-parameters of a two-part network showing incident waves ($a_1$ and $a_2$) and reflected waves ($b_1$ and $b_2$).

The unilateral gain $U$ is the forward power gain in a feedback amplifier with its reverse power gain set to zero by adjusting a lossless reciprocal feedback network around the HBT. This gain is defined as [34]

$$U = |s_{11}s_{22}s_{12}s_{21}|/[(1 - |s_{11}|^2)(1 - |s_{22}|^2)] \tag{2.78}$$

Another important figure of merit is the maximum oscillation frequency $f_{max}$, which is the frequency at which unilateral gain is unity (i.e., $U = 1$). This frequency is related to the cutoff frequency $f_T$ as

$$f_{max} \approx (1/2S)(f_T/2\pi r_B C_{jC})^{0.5} \tag{2.79}$$

where $S$ is the emitter stripe width, $r_B$ is the base resistance, and $C_{jC}$ is the base-collector junction capacitance.

S-parameters can also be simulated from circuit simulator SPICE [35] using the small-signal hybrid-$\pi$ equivalent circuit shown in Figure 2.23, which includes the intrinsic

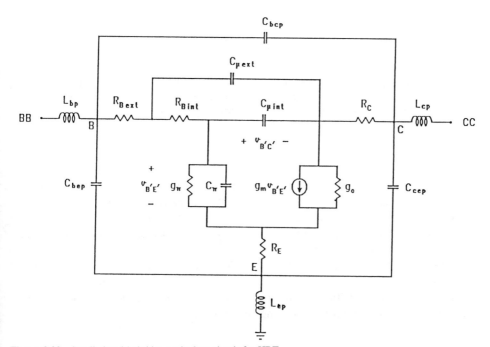

**Figure 2.23** Small-signal hybrid-p equivalent circuit for HBT.

(a)

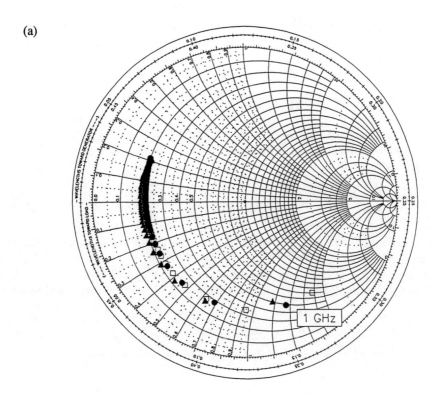

**Figure 2.24** (a) $S_{11}$, (b) $S_{12}$, (c) $S_{21}$, and (d) $S_{22}$ of the HBT obtained from measured (closed triangles), SPICE simulation without parasitic elements (open squares), and SPICE simulation with parasitic elements (closed circles) (*Source*: [36]. Reprinted with permission.).

elements as well as parasitic elements of the HBT. Since the parasitic inductances ($L_{ep}$, $L_{bp}$, and $L_{cp}$) and capacitances ($C_{cep}$, $C_{bep}$, and $C_{bcp}$) are difficult to model without considering a rigorous electromagnetic solution, their values are determined by estimating and fitting. Other model parameters can normally be determined from the knowledge of the material, geometry, and process [36]. Figures 2.24(a–d) illustrate the results of *s*-parameters for a typical HBT biased at $I_B = 800$ μA and $V_{CE} = 1$V. The measured results were obtained using a vector network analyzer. *S*-parameters simulated without parasitic elements and with parasitic elements are also given. Table 2.3 lists the model parameters used in SPICE simulation for obtaining such results.

(b)

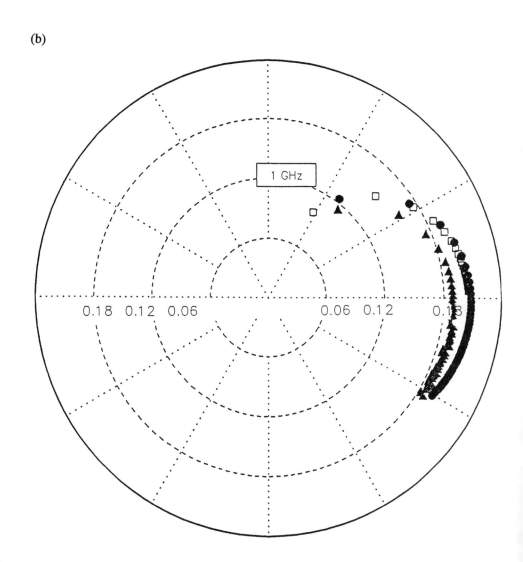

**Figure 2.24** (continued)

(c)

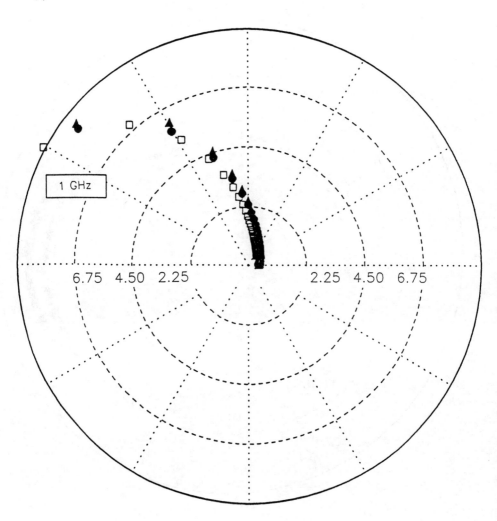

**Figure 2.24** (continued)

(d)

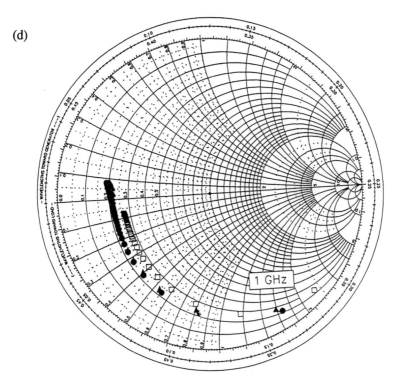

**Figure 2.24** (continued)

**Table 2.3**
Model Parameters Used in SPICE Simulation

| | |
|---|---|
| BF = 83.926 | TF = 0.95243 ps |
| BR = 0.6338 | TR = 0.12611 ns |
| NF = 1.1049 | CJE = 48.855 fF |
| NR = 1.0000 | CJC = 157.58 fF |
| NE = 1.7000 | MJE = 0.50 |
| NC = 1.7595 | MJC = 0.50 |
| ISE = $2.432 \times 10^{-19}$ A | VJE = 1.7018 V |
| ISC = $4.393 \times 10^{-16}$ A | VJC = 1.4107 V |
| IS = $1.111 \times 10^{-25}$ A | XCJC = 0.1590 |
| $R_E$ = 6.7896 Ω | $R_C$ = 2.200 Ω |
| $R_{Bint}$ = 0.7821 Ω | $R_{Bext}$ = 4.0557 Ω |
| $L_{ep}$ = 1.71 pH | $C_{cep}$ = 76.18 fF |
| $L_{bp}$ = 44.35 pH | $C_{bep}$ = 26.42 fF |
| $L_{cp}$ = 28.68 pH | $C_{bcp}$ = 76.01 fF |

Parameters are standard SPICE notations. See [35] for definitions.

# References

[1] Lundstrom, M. S., "An Eber-Moll Model for the Heterojunction Bipolar Transistor," *Solid-St. Electron.,* Vol. 29, 1986, p. 1173.

[2] Ryum, B. R., and I. M. Abdel-Motaleb, "A Gummel-Poon Model for Abrupt and Graded Heterojunction Bipolar Transistors," *Solid-St. Electron.,* Vol. 33, 1990, p. 869.

[3] Grinberg, A. A., M. S. Shur, R. J. Fischer, and H. Morkoc, "An Investigation of the Effect of Graded Layers and Tunneling on the Performance of AlGaAs/GaAs Heterojunction Bipolar Transistors," *IEEE Trans. Electron Devices,* Vol. ED-31, 1984, p. 1758.

[4] Tiwari, S., and D. J. Frank, "Analysis of the Operation of GaAlAs/GaAs HBT's," *IEEE Trans. Electron Devices,* Vol. 36, 1989, pp. 2105–2121.

[5] Liou, J. J., and J. S. Yuan, "Surface Recombination Current of AlGaAs/GaAs Heterojunction Bipolar Transistors," *Solid-St. Electron.,* Vol. 35, June 1992, p. 805.

[6] Henry, C. H., R. A. Logan, and F. R. Merritt, "The Effect of Surface Recombination on Current in $Al_xGa_{1-x}As$ Heterojunctions," *J. Appl. Phys.,* Vol. 49, 1978, pp. 3530–3542.

[7] Hiraoka, Y. S., and J. Yoshida, "Two-Dimensional Analysis of the Surface Recombination Effect on Current Gain for GaAlAs/GaAs HBT's," *IEEE Trans. Electron Devices,* Vol. 35, 1988, pp. 857–862.

[8] Wieder, H. H., "Surface Fermi Level of III–V Compound Semiconductor-Dielectric Interfaces," *Surface Sci.,* Vol. 132, 1983, pp. 390–405.

[9] Tiwari, S., D. J. Frank, and S. L. Wright, "Surface Recombination in AlGaAs/GaAs Heterostructure Bipolar Transistors," *J. Appl. Phys.,* Vol. 64, 1988, p. 5009.

[10] Liu, W., and J. S. Harris, Jr., "Diode Ideality Factor for Surface Recombination Current in AlGaAs/GaAs Heterojunction Bipolar Transistors," *IEEE Trans. Electron Devices,* Vol. 39, 1992, p. 2726.

[11] Shur, M. S., "Recombination Current in Forward-Biased p-n Junctions," *IEEE Trans. Electron Devices,* Vol. 35, 1988, pp. 1564–1565.

[12] Parikh, C. D., and F. A. Lindholm, "Space-Charge Region Recombination in Heterojunction Bipolar Transistors," *IEEE Trans. Electron Devices,* Vol. 39, 1992, p. 2197.

[13] Bowler, D. L., and F. A. Lindholm, "High Current Regimes in Transistor Collector Regions," *IEEE Trans. Electron Devices,* Vol. ED-20, 1973, p. 257.

[14] Kirk, Jr., C. T., "A Theory of Transistor Cutoff Frequency Falloff at High Current Densities," *IRE Trans. Electron Devices,* Vol. ED-9, 1962, p. 164.

[15] Asbeck, P. M., et al., "Heterojunction Bipolar Transistors for Ultra High Speed Digital and Analog Applications," *IEDM Tech. Digest,* 1988.

[16] Rockett, P. I., "Monte Carlo Study of the Influence of Collector Region Velocity Overshoot on the High-Frequency Performance of AlGaAs/GaAs Heterojunction Bipolar Transistors," *IEEE Trans. Electron Devices,* Vol. ED-35, October 1988, pp. 1573–1579.

[17] Laux, S. E., and W. Lee, "Collector Signal Delay in the Presence of Velocity Overshoot," *IEEE Electron Device Lett.,* Vol. 11, 1990, p. 174.

[18] Ishibashi, T., "Influence of Electron Velocity Overshoot on Collector Transit Time of HBT's," *IEEE Trans. Electron Devices,* Vol. 37, 1990, p. 2103.

[19] Liou, J. J., and H. Shakouri, "Collector Signal Delay Time and Collector Transit Time of HBT's Including Velocity Overshoot," *Solid-St. Electron.,* Vol. 35, 1992, p. 15.

[20] Katoh, R., and M. Kurata, "Self-Consistent Particle Simulation for AlGaAs/GaAs HBT's Under High Bias Conditions," *IEEE Trans. Electron Devices,* Vol. 36, 1989, p. 2122.

[21] Dutton, R. W., "Bipolar Transistor Modeling of Avalanche Generation for Computer Circuit Simulation," *IEEE Trans. Electron Devices,* Vol. ED-22, June 1975, pp. 334–338.

[22] Divekar, D. A., and R. E. Lovelace, "Modeling of Avalanche Generation Current of Bipolar Junction Transistors for Computer Circuit Simulation," *IEEE Trans. Computer-Aided Design,* Vol. CAD-1, July 1982, pp. 112–116.

[23] Hebert, F., and D. J. Roulston, "Modeling of Narrow-Base Bipolar Transistors Including Variable-Base-Charge and Avalanche Effects," *IEEE Trans. Electron Devices,* Vol. ED-34, November 1987, pp. 2323–2328.
[24] Kloosterman, W. J., and H. C. De Graaff, "Avalanche Multiplication in a Bipolar Transistor Model for Circuit Simulation," *IEEE Bipolar Transistor Circuits & Technology Meeting,* Minneapolis, MN, 1988.
[25] Liou, J. J., and J. S. Yuan, "Modeling the Reverse Base Current Phenomenon due to Avalanche Effect in Advanced Bipolar Transistors," *IEEE Trans. Electron Devices,* Vol. 37, 1990, p. 2274.
[26] Lu, P.-F., and T.-C. Chen, "Collector-Base Junction Avalanche Effects in Advanced Double-Poly Self-Aligned Bipolar Transistors," *IEEE Trans. Electron Devices,* Vol. 36, 1989, p. 1182.
[27] Liou, J. J., and J. S. Yuan, "An Avalanche Multiplication Model for Bipolar Transistors," *Solid-St. Electron.,* Vol. 33, January 1990, pp. 35–38.
[28] Yang, E. S., *Microelectronic Devices,* New York: McGraw-Hill, 1988.
[29] Poon, P. S., and J. C. Meckwood, "Modeling of Avalanche Effects in Integral Charge Model," *IEEE Trans. Electron Devices,* Vol. ED-19, January 1972, pp. 90–97.
[30] Selberherr, S., *Analysis and Simulation of Semiconductor Devices,* New York: Spring-Verlag, 1984, p. 111.
[31] Muller, R. S., and T. I. Kamins, *Device Electronics for Integrated Circuits,* 2nd ed., New York: Wiley, 1986.
[32] Liou, J. J., "Calculation of the Base Current Components and Determination of Their Relative Importance in AlGaAs/GaAs and lnAlAs/lnGaAs Heterojunction Bipolar Transistors," *J. Appl. Phys.,* Vol. 69, 1991, p. 3328.
[33] Kurokawa, K., "Power Waves and the Scattering Matrix," *IEEE Trans. Microwave Theory and Tech.,* Vol. MTT-13, 1965, p. 194.
[34] Liao, S. Y., *Microwave Devices and Circuits,* 2nd edition, Englewood Cliffs, NJ: Prentice-Hall, 1985.
[35] Antognetti, P., and G. Massobrio, *Semiconductor Device Modeling with SPICE,* New York: McGraw-Hill, 1988.
[36] Fellows, J. A., V. M. Bright, and T. J. Jenkins, "An Accurate Physics-Based Heterojunction Bipolar Transistor Model for SPICE-Assisted Integrated Circuit Design," *IEE Proc.,* to appear.

# Chapter 3
# HBTs With Enhanced Structures

The performance of abrupt AlGaAs/GaAs HBTs can be improved with enhanced structures such as graded base, graded heterojunction, and setback layer. A graded base gives rise to an aiding built-in electric field in the quasi-neutral base, which results in a reduction in the minority-carrier transit time in the base and thus an increase in the common-emitter current gain $\beta$ and the cutoff frequency $f_T$ of the HBT [1–4]. Also, as mentioned in the beginning of the first chapter, removing the conduction band discontinuity (spike) at the abrupt heterointerface can improve the emitter injection efficiency. Two approaches have been widely used to achieve this. One approach is to insert a graded layer before the heterointerface in which the Al composition is decreased gradually from 0.3 to 0.0. The other is to insert an undoped GaAs layer (setback layer) after the heterointerface.

In addition to these intrinsic enhanced structures, extrinsic enhanced structures, such as the passivation emitter ledge [5] and the proton-implanted collector [6], are used frequently to improve HBT performance. The former can reduce the electron-hole recombination at the extrinsic base surface and thus increase the current gain, and the latter can reduce the extrinsic base-collector junction capacitance and subsequently increase the cutoff frequency.

In this chapter, physics-based models for the AlGaAs/GaAs HBT with enhanced structures are developed and the characteristics of such devices are discussed.

## 3.1 BASE GRADING

The HBT performance can be further improved by utilizing a graded base structure. Figure 3.1 illustrates qualitatively the energy band diagram of the graded base HBT that has an aiding field $E_{aid}$ for electrons. Note that $E_{aid} = -\xi_B$, where $\xi_B$ is the base built-in field. $E_{aid}$ can be obtained by varying linearly the Al mole fraction $x'$ of the material in the base, which varies the electron affinity $\chi$ and thus the conduction-band edge in the base. In general, using the coordinates specified in Figure 1.2,

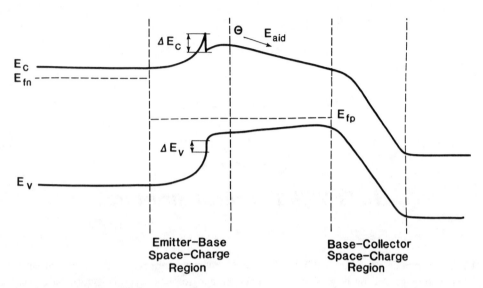

**Figure 3.1** Energy band diagram of a graded-base HBT.

$$|\xi_B| = [\chi_B(X_3) - \chi_B(X_2)]/(X_3 - X_2) \approx [\chi_B(X_{jC}) - \chi_B(0)]/X_{jC} \quad (3.1)$$

where $X_2$ and $X_3$ are the edges of the quasi-neutral base close to the emitter and collector junctions, respectively; 0 and $X_{jC}$ are the emitter-base and base-collector metallurgical junctions; and $\chi_B$ is the electron affinity in the base (in eV). For the $Al_xGa_{1-x}As$ base under study, $\chi_B = 4.07 - 1.1x'$. For instance, if $x'$ varies linearly from 0.3 at $x = 0$ to 0.0 at $x = X_{jC}$, then $|\xi_B| = (4.07 - 3.74)/X_{jC}$.

### 3.1.1 Effect of Base Grading on Current Gain

The electron current density $J_n$ in the base including $\xi_B$ can be expressed as

$$J_n = -qnv_d + q\mu_n V_T dn/dx \quad (3.2)$$

where $v_d$ is the drift velocity. Based on Monte Carlo simulation results reported by Rockett (Figure 3.2) [7] for an HBT with a base grading of 0.12 to 0.0 and a base width of 0.25 μm (this represents an $\xi_B \approx 5$ kV/cm), we empirically model $v_d$ as $v_d \approx 10^7$ cm/s when $\xi_B = 0$; if we assume $v_d$ depends linearly upon $\xi_B$, then $v_d(\xi_B) \approx 10^7 + (1 \times 10^7/5000)|\xi_B|$ in the base of $AlGaAs/Al_xGa_{1-x}As$ HBTs. For an HBT with $x' = 0.2$ to 0.0 in the base and a base width of 1000 Å, the empirical $v_d$ model predicts $v_d \approx 5 \times 10^7$ cm/s, which is in reasonable agreement with $v_d$ simulated by Horio et al. [8] ($\approx 4 \times 10^7$ cm/s) but is larger than that computed by Katoh et al. [9] ($\approx 2.5 \times 10^7$ cm/s. Also

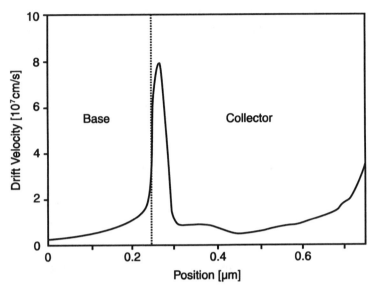

**Figure 3.2** Drift velocity obtained from Monte Carlo simulation for a typical HBT with a 0.25 mm base width and no base grading (*after* [7]).

$\mu_n \approx 943 + \{7057/[1 + (N_B/2.8 \times 10^{16})^{0.753}]\}$ cm$^2$/V-sec, where $N_B$ is the base doping density, without built-in field and $\mu_n = v_d(\xi_B)/|\xi_B|$ otherwise [10].

Since the base region of the HBT is very thin, the recombination current in the base is negligible and $J_n$ can be assumed constant throughout the base. Thus taking a derivative on both sides of (3.2) with respect to position yields

$$0 = -v_d dn/dx + \mu_n V_T d^2n/dx^2 \quad (3.3)$$

Solving for $n$ from (3.3) yields

$$n(x) = A + B \exp[(v_d/\mu_n V_T)x] \quad (3.4)$$

Using the boundary conditions $n(X_2)$ and $n(X_3)$, the constants $A$ and $B$ can be solved and

$$n(x) = \frac{-n(X_2)\exp(aX_3) + n(X_3)\exp(aX_2) + [n(X_2) - n(X_3)]\exp(ax)}{\exp(aX_2) - \exp(aX_3)} \quad (3.5)$$

where $X_2$ and $X_3$ are the space-charge region edges in the base and $a = v_d/\mu_n V_T$. The results of $n(x)$ in the base for Al$_{0.3}$Ga$_{0.7}$As/Al$_x$Ga$_{1-x}$As/GaAs HBTs that have $N_E = 2 \times 10^{17}$ cm$^{-3}$, $N_B = 10^{18}$ cm$^{-3}$, $N_C = 5 \times 10^{16}$ cm$^{-3}$, emitter layer thickness of 0.2 μm, and base layer thickness of 0.2 μm are plotted in Figure 3.3 for $V_{BE} = 1.0$V and $V_{CB} = 2.0$V. The composition $x'$ in the base appears in the upper-right corner of the figure; the first value represents $x'$ at $x = X_2$ and the second represents $x'$ at $x = X_3$. It should be pointed out that the profiles of $n(x)$ for the devices with $|\xi_B| = 0$ ($x' = 0.0$ to $0.0$ and $x' = 0.1$ to $0.1$) are closer to a linear function with respect to $x$ than that for the devices with $|\xi_B| > 0$ ($x' = 0.1$ to $0.0$ and $x' = 0.2$ to $0.0$). This trend agrees with the conventional theory, which assumes $\xi_B = 0$ and employs the diffusion-current-only approximation, that $n(x)$ in the base is linear with respect to $x$. Putting (3.5) into (3.2) yields

$$J_C = J_n(X_3) = -qn(X_3)v_d + q\mu_n V_T dn/dx|_{X_3}$$

$$= -qn(X_3)v_d + (q\mu_n V_T)\frac{a[n(X_2) - n(X_3)]\exp(aX_3)}{\exp(aX_2) - \exp(aX_3)} \quad (3.6)$$

Note that $n(X_3) \approx n_0(X_3)$ and the expression for $n(X_2)$ has been given in the previous section.

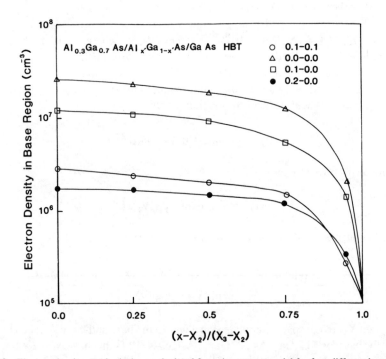

**Figure 3.3** Electron density $n(x)$ in the base calculated from the present model for four different base gradings.

Figure 3.4 compares the dc current gain β versus $J_C$ calculated from the present model and obtained from Monte Carlo simulation for the same devices used in Figure 3.3. Similar trends are obtained. The present model is also compared with experimental data, as shown in Figure 3.5, and good agreement is found.

Figure 3.6 shows β calculated from the present model for an AlAs/Al$_x$Ga$_{1-x}$As/GaAs HBT for four different base gradings. It is shown that a base grading of $x' = 0.2$ to 0.0 yields a smaller β than that of $x' = 0.1$ to 0.0, contrasting with the conventional concept that increasing $\xi_B$ will always increase β. This is because the heterojunction barrier heights depend strongly on $x'$ ($x = X_2$); thus although a base grading of $x' = 0.2$ to 0.0 will yield a larger $\xi_B$ and enhance the charge transport in the base, it nonetheless will reduce the valence-band barrier height, which consequently increases $J_B$ and decreases β. The benefit of $\xi_B$ on β can be better appreciated by comparing devices that have the same $x'$ ($x = X_2$), as evidenced by the results in Figure 3.6 in which a device with $x' = 0.1$ to 0.0 possesses a larger β than a device with $x' = 0.1$ to 0.1.

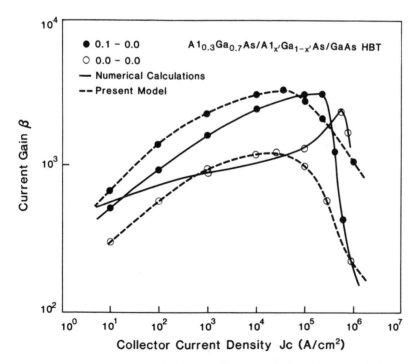

**Figure 3.4** Comparison of β versus $J_C$ calculated from the present model (dashed lines) and obtained from Monte Carlo simulation (solid lines).

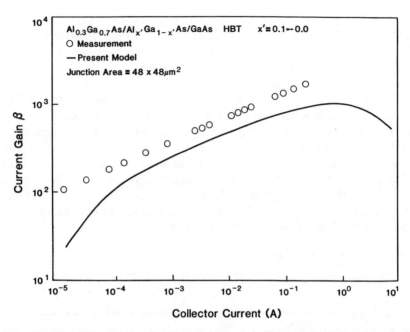

**Figure 3.5** Comparison of β calculated from the present model (solid lines) and obtained from measurement (circles) reported in [1].

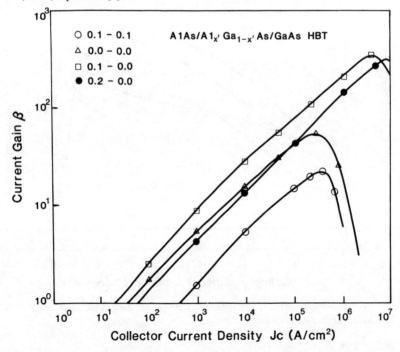

**Figure 3.6** Current gain β versus $J_C$ calculated from the present model for four different base gradings.

## 3.1.2 Effect of Base Grading on Early Voltage

The Early voltage $V_A$ is an important parameter for bipolar transistors; it describes the dc collector current-voltage characteristics as well as the small-signal output resistance of the device under the active operation. The Early voltage of an HBT is much larger than that of a Si homojunction *bipolar transistor* (BJT) because the base doping concentration in the HBT is higher than that in the Si BJT, which subsequently reduces the HBT base width modulation caused by the change of the voltage applied to the base-collector junction [11].

Here we develop an Early voltage model including the effect of base grading. Consider the same HBT (N/p/n $Al_{0.3}Ga_{0.7}As/Al_xGa_{1-x'}As/GaAs$) as used in the previous section. The Early voltage $V_A$ is defined related to the slope of $J_C$ versus the collector-emitter voltage $V_{CE}$ in the active mode (Figure 3.7):

$$dJ_C/dV_{CE}|_{V_{BE}=const} = \partial J_C/\partial V_{CB} = J_C/(V_A + V_{CE}) \approx J_C/V_A \quad (3.7)$$

Thus

$$V_A = J_C(\partial J_C/\partial V_{CB})^{-1} \quad (3.8)$$

where the equation for $J_C$ is given in the previous section and

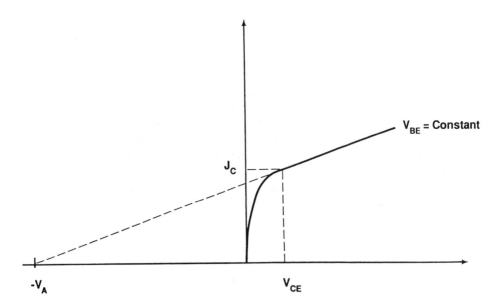

**Figure 3.7** The current-voltage characteristics illustrating the Early voltage $V_A$.

$$\partial J_C/\partial V_{CB} = qn(X_3)v_d/V_T + q\mu_n V_T\{a^2n(X_2)\exp(aX_3)(dX_3/dV_{CB})[\exp(aX_2) - \exp(aX_3)]^{-1}$$

$$+ a^2n(X_2)\exp(aX_3)[\exp(aX_2) - \exp(aX_3)]^{-2}\exp(aX_3)(dX_3/dV_{CB})\} \quad (3.9)$$

Here

$$dX_3/dV_{CB} = -[0.5N_C/(N_C + N_B)](V_{bi,BC} + V_{CB})^{-0.5}[2(\varepsilon_C/q)(1/N_C + 1/N_B)]^{0.5} \quad (3.10)$$

To illustrate the effect of the base grading on the Early voltage, we consider a typical $Al_{0.3}Ga_{0.7}As/Al_{x'}Ga_{1-x'}As/GaAs$ HBT with the following device make-up: $N_E = 10^{18}$ cm$^{-3}$, $N_B = 10^{19}$ cm$^{-3}$, $N_C = 5 \times 10^{16}$ cm$^{-3}$, emitter layer thickness of 1500 Å, and base layer thickness of 1000 Å. Figure 3.8(a) plots $V_A$ versus $V_{CB}$ for base grading structures of $x' = 0.0$ to 0.0 (no grading), $x' = 0.1$ to 0.0, $x' = 0.2$ to 0.0, and $x' = 0.3$ to 0.0 calculated at $V_{BE} = 1.4$V, which is the voltage at which most AlGaAs/GaAs HBTs are design to operate. The results suggest that $V_A$ increases significantly as the base grading is increased. Also, $V_A$ depends more strongly on $V_{CB}$ when $x'$ is larger. Figure 3.8(b) illustrates the dependence of $V_A$ on $V_{BE}$ for the four base grading structures, which shows that $V_A$ increases only slightly when $x'$ is small and increases more notably when $x'$ is large. It should be pointed out that the Early voltage in general depends on $V_{CB}$ as well as on $V_{BE}$ because the quasi-neutral base width is modulated by both voltages. While it seems that a large base grading like $x' = 0.3$ to 0.0 should be used for the $V_A$ point of view, such a base grading will nonetheless degrade other HBT performances, like the dc current gain β, as shown in the previous section.

## 3.2 HBTs WITH A SETBACK LAYER

Inserting a thin layer of intrinsic GaAs (called the setback layer or spacer) between the emitter and base can alter the barrier potentials on both sides of the heterojunction, thus reducing the importance of thermionic and tunneling mechanisms on the free-carrier transport across the heterojunction [12]. An additional advantage of the setback layer is that it can prevent impurity out-diffusion from the heavily doped base to emitter.

Focus on the N/p $Al_{0.3}Ga_{0.7}As/GaAs$ heterojunction in which an undoped GaAs layer is inserted between the emitter and base layers (between $x = 0$ and $x = W_i$), as shown in Figure 3.9. Solving the Poisson equation including the effects of the setback layer and assuming the setback layer is fully depleted, the electric fields $\xi(x)$ in the space-charge layer are given by [12]

$$\xi_1(x) = (qN_E/\varepsilon_E)(x + X_1) \quad -X_1 \leq x \leq 0 \quad (3.11)$$

$$\xi_s(x) = (qN_B/\varepsilon_B)(X_2 - W_i) \quad 0 \leq x \leq W_i \quad (3.12)$$

$$\xi_2(x) = (qN_B/\varepsilon_B)(X_2 - x) \quad W_i \leq x \leq X_2 \quad (3.13)$$

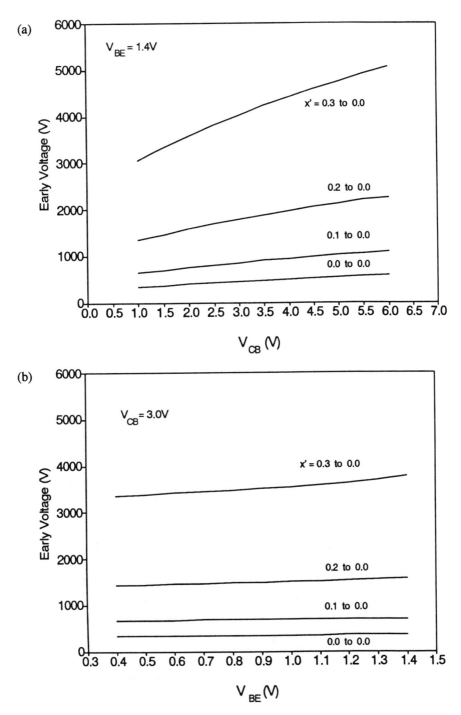

**Figure 3.8** Early voltage versus (a) the base-collector voltage and (b) the emitter-base voltage for four different base grading structures.

where $W_i$ is the setback layer thickness and $X_1$ and $X_2$ are the thicknesses on the emitter and base sides of the space-charge region, respectively (Figure 3.9). Integrating (3.11) to (3.13) over their boundaries, the barrier potentials $V_{B1}$ and $V_{B2}$ on the emitter and base sides, respectively, can be derived as

$$V_{B1} = qN_E X_1^2/(2\varepsilon_E) \tag{3.14}$$

$$V_{B2} = qN_B(X_2 - W_i)W_i/\varepsilon_B + qN_B(X_2 - W_i)^2/(2\varepsilon_B) \tag{3.15}$$

Note that the first term on the right-hand side of (3.15) is the barrier potential associated with the setback layer.

Using the conditions that the electrostatic potential must be continuous at $x = 0$ and that the charge neutrality exists over the space-charge region, the space-charge region thicknesses $X_1$ and $X_2$ can be solved as [12]

$$X_1 = -[W_i\varepsilon_E N_B/(\varepsilon_E N_E + \varepsilon_B N_B)]$$
$$+ \{[W_i\varepsilon_E N_B/(\varepsilon_E N_E + \varepsilon_B N_B)]^2 + 2\varepsilon_E\varepsilon_B(V_{bi} - V_{BE})N_B/[qN_E(\varepsilon_E N_E + \varepsilon_B N_B)]\}^{0.5} \tag{3.16}$$

$$X_2 = X_1 N_E/N_B + W_i \tag{3.17}$$

Note that if $W_i = 0$, then (3.16) reduces to

$$X_1 = \{2\varepsilon_E\varepsilon_B(V_{bi} - V_{BE})N_B/[qN_E(\varepsilon_E N_E + \varepsilon_B N_B)]\}^{0.5} \tag{3.18}$$

which is the space-charge region thickness model for an abrupt heterojunction.

For illustrations we consider a typical N/p abrupt heterojunction having $N_E = 5 \times 10^{17}$ cm$^{-3}$, $N_B = 10^{19}$ cm$^{-3}$, and three different setback layer thicknesses $W_i = 0$, 100 Å, and 200 Å. Figure 3.10 shows the space-charge region thicknesses $X_1$ and $X_2$ versus the

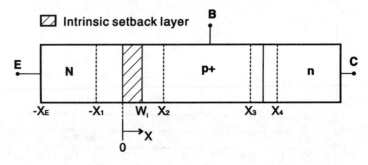

**Figure 3.9** Schematic illustration of the HBT structure including a setback layer.

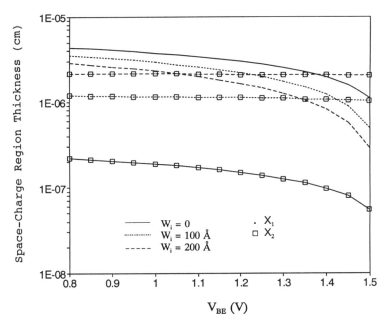

**Figure 3.10** Space-charge region thicknesses versus $V_{BE}$ calculated for three different setback layer thicknesses.

applied voltage $V_{BE}$. In general, $X_1$ is larger than $X_2$ because the doping concentration in the base is higher than that in the emitter, except for the cases of $W_i > 0$ and $V_{BE}$ is large. In these cases, $X_2 \approx W_i$ and can be larger than $X_1$ if $V_{BE}$ is large. The barrier potential $V_{B2}$ on the base-side of the junction is also increased due to the presence of the setback layer, as evidenced by the results in Figure 3.11. On the other hand, the barrier potential $V_{B1}$ on the emitter-side decreases as $W_i$ increases.

### 3.2.1 Collector Current

Following the thermionic-field-diffusion approach developed by Grinberg et al. [13], the electron current density $J_n$ across the heterointerface ($x = 0$) is the difference between two opposing fluxes

$$J_n(0) = qv_n\gamma_n[n(0^-) - n(0^+)\exp(-\Delta E_C/kT)] \quad (3.19)$$

where $v_n$ is the electron thermal velocity, $\gamma_n$ is the electron tunneling coefficient (see previous section), $n$ is the electron concentration, and $\Delta E_C \approx 0.6\Delta E_G$ is the conduction-and discontinuity (spike). In (3.19),

$$n(0^-) = N_E\exp(-V_{B1}/V_T) \quad \text{and} \quad n(0^+) = n(X_2)\exp(V_{B2}/V_T) \quad (3.20)$$

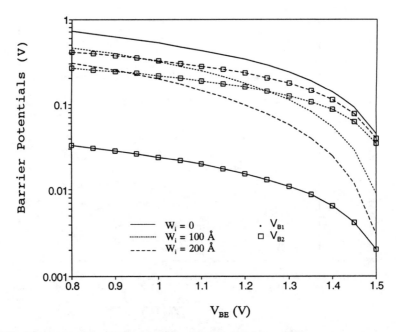

**Figure 3.11** Barrier potentials calculated for three different setback layer thicknesses.

At this point, the only unknown parameter is $n(X_2)$, which can be solved using the relation

$$J_n(0) = J_{SCRI} + J_{SCRB} + J_n(X_2) \tag{3.21}$$

where $J_{SCRI}$ is the recombination current density in the setback layer ($x = 0$ and $x = W_i$), $J_{SCRB}$ is recombination current density in the space-charge layer associated with the base layer ($x = W_i$ and $x = X_2$) (the models for $J_{SCRI}$ and $J_{SCRB}$ will be developed in later), and $J_n(X_2)$ is the diffusion-only current in the *quasi-neutral base* (QNB). For a very thin base, the base transport factor $\alpha_B$ approaches unity and the collector current density $J_C$ is given by

$$J_C = J_n(X_2) = qD_n n(X_2)/(W_B + \Delta W_B + D_n/v_s) \tag{3.22}$$

where $\Delta W_B$ is the current-induced base pushout. Incorporating the above equations and solving for $n(X_2)$, we obtain

$$n(X_2) = [qv_n\gamma_n N_E \exp(-V_{B1}/V_T) - J_{SCRI} - J_{SCRB}]/\zeta \tag{3.23}$$

where

$$\zeta = qD_n/(W_B + \Delta W_B + D_n/v_s) + qv_n\gamma_n \exp[(V_{B2} - \Delta E_C/q)/V_T] \tag{3.24}$$

## 3.2.2 Base Current

The base current density $J_B$ model for the abrupt HBT (discussed in Chapter 2) still applies here provided the total space-charge region recombination current density $J_{SCR}$ is modified to account for the electron-hole recombination in the setback layer. $J_{SCR}$ now consists of recombination current densities occurring in the emitter-side of the space-charge layer ($J_{SCRE}$), in the intrinsic layer ($J_{SCRI}$), and in the base-side of the space-charge layer ($J_{SCRB}$). Thus

$$J_{SCRB} = J_{SCRE} + J_{SCRI} + J_{SCRB} \tag{3.25}$$

$$= q \int_{-X_1}^{0} U_{SRH,E}\, dx + q \int_{0}^{W_i} U_{SRH,I}\, dx + q \int_{W_i}^{X_2} U_{SRH,B}\, dx \tag{3.26}$$

where $U_{SRH}$ is the Shockley-Read-Hall recombination rate. Assuming a single-level trap located at the middle of the bandgap yields

$$U_{SRH,E} = 0.5\sigma v_n N_t n_{iE} \exp(V_{BE}/2V_T) \tag{3.27}$$

$$U_{SRH,I} = U_{SRH,B} = 0.5\sigma v_n N_t n_{iB} \exp(V_{BE}/2V_T) \tag{3.28}$$

For illustrations we consider a typical HBT with three different setback layer thicknesses $W_i = 0$, 100 Å, and 200 Å. The collector current densities calculated from the present model for three $W_i$ are shown in Figure 3.12. Also included are the results obtained from a numerical model [14] that solves numerically the Poisson and continuity equations. Good agreement is found between the present and numerical models. Clearly, the setback layer improves the emitter injection efficiency, particularly when $V_{BE}$ is relatively low. Also note that there is a smaller increase in $J_C$ when $W_i$ is increased from 100 Å to 200 Å than from 0 to 100 Å. We also calculated the collector current density $J_C'$ without including the tunneling mechanism. The results of $J_C'/J_C$ are illustrated in Figure 3.13, which indicate that the tunneling current becomes much less important when $W_i$ is increased, stemming from the smaller tunneling probability caused by the decreased barrier potential $V_{B1}$. Our finding agrees with numerical results presented in [15], which suggested that the conventional drift-diffusion model becomes applicable if a relatively thick setback layer is used.

The setback layer increases the base current density of the HBT as well, as shown in Figure 3.14. To investigate the origin of this increase, we calculate the four components of $J_B$ for $W_i = 0$, which are given in Figure 3.15. The results show that $J_{SCR}$ and $J_{RB}$ are the dominant components of $J_B$ at low and high $V_{BE}$, respectively. Our calculations also suggest that both $J_{SCR}$ and $J_{RB}$ increase with increasing $W_i$, which then contributes to the increased $J_B$ found in Figure 3.14.

In Figure 3.16, the dc current gains $\beta$ versus $J_C$ that are calculated from the present

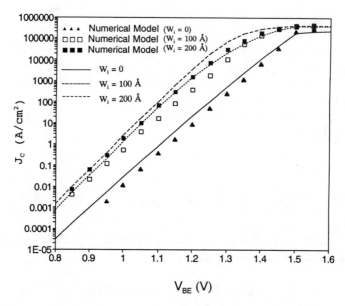

**Figure 3.12** Collector current densities versus $V_{BE}$ calculated from the present model and from the numerical model [14].

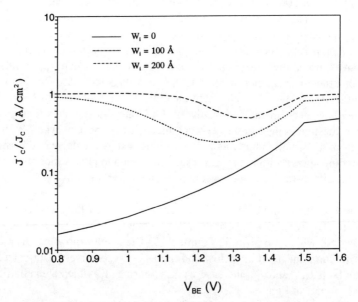

**Figure 3.13** Comparison of the collector current density without tunneling ($J'_C$) and with tunneling ($J_C$).

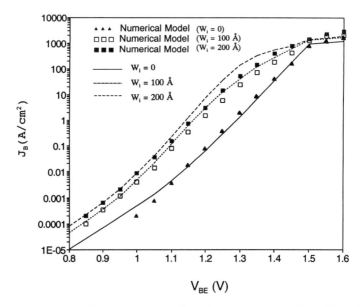

**Figure 3.14** Base current densities versus $V_{BE}$ calculated from the present model and from the numerical model [14].

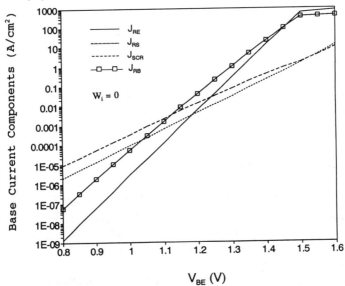

**Figure 3.15** Four base current density components calculated as a function of $V_{BE}$. $J_{RE}$ is the injection current density from the base to emitter; $J_{RS}$ is the total surface recombination current density including $J_{BS}$ and $J_{ES}$; $J_{SCR}$ is the recombination current density in the space-charge region; and $J_{RB}$ is the recombination current in the quasi-neutral base.

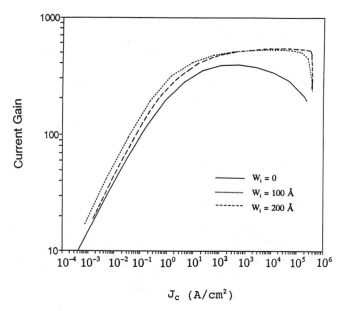

**Figure 3.16** Comparison of the dc current gains versus $J_C$ calculated from the present model for three different setback layer thicknesses.

model for three $W_i$ are compared. For the device make-up considered and the parameters used, our calculations suggest that inserting a setback layer between the emitter and base in general improves the current gain. A 200-Å setback layer nonetheless yields a slightly smaller $\beta$ than a 100-Å setback layer at the low and medium current levels, and similar current gains are obtained for $W_i = 100$ Å and 200 Å at the high current region. As a result, using a layer thicker than 100 Å actually degrades the HBT performance if the bias condition is relatively low.

## 3.3 HBTs WITH A GRADED LAYER

A graded layer inserted between the emitter and the base (Figure 3.17) is also often used to improve the free-carrier injection efficiency. Such a layer, in which the Al mole fraction is decreased linearly from 0.3 to 0.0 and normally has a thickness between 100 Å and 300 Å, can effectively remove the spike and thus make the thermionic and tunneling mechanisms less important.

The graded HBT case is essentially similar to the homojunction bipolar transistor except that forces acting on the electrons and holes must include energy-gap variations in addition to the electric field [16]. Marty et al. [17] first presented a detailed theory for the abrupt HBT, but, as pointed out recently by Grinberg and Luryi [18], their model really applies only to the graded HBT because the model implicitly relies on the continuity of

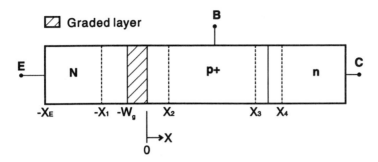

**Figure 3.17** Schematic illustration of the HBT structure including a graded layer.

the quasi-Fermi level at the emitter-base heterojunction, which is an assumption that is valid only for the graded HBT.

The dielectric permittivity $\varepsilon_g(x)$ in the linearly graded layer is

$$\varepsilon_g(x) = -[(\varepsilon_E - \varepsilon_B)/W_g]x + \varepsilon_B \tag{3.29}$$

where $W_g$ is the graded layer thickness (Figure 3.17). For the $Al_{0.3}Ga_{0.7}As/GaAs$ junction, $\varepsilon_E = 12.2\varepsilon_0$ and $\varepsilon_B = 13.1\varepsilon_0$ [19] and $\varepsilon_g(x) = (0.93\varepsilon_0/W_g)x + \varepsilon_B$.

For the graded layer that has a position-dependent permittivity, the one-dimensional Poisson equation is

$$d^2V/dx^2 = -\{\rho/\varepsilon_g(x) + (dV/dx)[d\varepsilon_g(x)/dx]\varepsilon_g(x)\}$$

$$= -\{\rho/\varepsilon_g(x) - (dV/dx)[(\varepsilon_E - \varepsilon_B)/W_g]/\varepsilon_g(x)\} \tag{3.30}$$

where $V$ is the electrostatic potential. For a typical emitter doping concentration of $5 \times 10^{17}$ cm$^{-3}$ and graded layer thickness of 200 Å, the second term on the right-hand side of (3.30) is negligibly small compared to the first term unless the electric field $dV/dx$ exceeds $5 \times 10^5$ V/cm, which is unlikely in the forward-biased heterojunction (forward-active mode) under study. Assuming the second term is zero and employing the depletion approximation, the electric field $\xi_g(x) = -dV/dx$ in the graded layer can be derived as [20]

$$\xi_g(x) = \xi_g(0) + [qN_EW_g/(0.93\varepsilon_0)]\ln[1 + 0.93\varepsilon_0 x/(W_g\varepsilon_B)] \quad \text{for } -W_g \leq x \leq 0 \tag{3.31}$$

Similarly, the electric fields $\xi_2(x)$ and $\xi_1(x)$ on the base and emitter sides of the space-charge region can be derived as

$$\xi_2(x) = (qN_B/\varepsilon_B)(X_2 - x) \quad \text{for } 0 \leq x \leq X_2 \tag{3.32}$$

$$\xi_1(x) = (qN_E/\varepsilon_E)(X_1 + x) \quad \text{for } -X_1 \leq x \leq -W_g \tag{3.33}$$

Since the flux density at $x = 0$ is continuous, $\xi_g(0)$ needed in (3.31) can be readily obtained from (3.32) as $\xi_g(0) = qN_B X_2/\varepsilon_B$.

Choose $x = -X_1$ as the reference. The corresponding position-dependent electrostatic potentials $V(x)$ on the emitter side of the junction can be obtained by integrating $\xi(x)$ over their boundaries as

$$V_1(x) = \int_{-X_1}^{x} \xi_1(x) \, dx \tag{3.34}$$

$$V_g(x) = V_1(-W_g) + \int_{-W_g}^{x} \xi_g(x) \, dx \tag{3.35}$$

Since $0.93\varepsilon_0 x/(W_g \varepsilon_B)$ is smaller than unity for $0 < x < W_g$, we can use the series expansion for the logarithm function in (3.31) and truncate the higher order terms, yielding

$$\ln(1 + Z) \approx Z - Z^2/2 \tag{3.36}$$

where $Z = 0.93\varepsilon_0 x/(W_g \varepsilon_B)$. Combining (3.31) to (3.36) yields [20]

$$V_1(x) = 0.5(qN_E/\varepsilon_E)(X_1 + x)^2 \quad \text{for } -X_1 \leq x \leq -W_g \tag{3.37}$$

$$V_g(x) = 0.5(qN_E/\varepsilon_E)(X_1 - W_g)^2 + (qN_B X_2/\varepsilon_B)(x + W_g) + 0.5(qN_E/\varepsilon_E)(x^2 - W_g^2)$$
$$- (qN_E/6)[0.93\varepsilon_0/(W_g \varepsilon_B^2)](x^3 + W_g^3) \quad \text{for } -W_g \leq x \leq 0 \tag{3.38}$$

Using the same approach and noting that the total electrostatic potential across the space-charge region is $V_{bi} - V_{BE}$, the electrostatic potential $V_2(x)$ on the base side of the junction is given by

$$V_2(x) = V_{bi} - V_{BE} - 0.5(qN_B/\varepsilon_B)(X_2 - x)^2 \quad \text{for } 0 \leq x \leq X_2 \tag{3.39}$$

The space-charge region thickness $X_2$ can be solved from (3.38) and (3.39) using the condition that the potential is continuous at $x = 0$ ($V_g(0) = V_2(0)$) [20]:

$$X_2 = -0.5A_2/A_1 - 0.5(A_2^2 - 4A_1 A_3)^{0.5}/A_1 \tag{3.40}$$

where

$$A_1 = -0.5qN_B/\varepsilon_B - 0.5qN_B^2/(\varepsilon_E N_E)$$

$$A_2 = qN_BW_g/\varepsilon_E - qN_BW_g/\varepsilon_B$$

$$A_3 = V_{bi} - V_{BE} + 0.5qN_EW_g^2/\varepsilon_B - 0.5qN_EW_g^2/\varepsilon_E + qN_EW_g^2(0.93\varepsilon_0/\varepsilon_B^2)/6$$

Since the flux density is continuous at $x = -W_g$, the relation between $X_1$ and $X_2$ can be derived by equating (3.31) and (3.33) at $x = -W_g$.

For the case of $W_g = 0$ (abrupt junction), $A_2 = 0$, $A_3 = V_{bi} - V_{BE}$, and (3.40) reduces to

$$X_2 = \{2(V_{bi} - V_{BE})\varepsilon_E\varepsilon_B N_E/[qN_B(\varepsilon_E N_E + \varepsilon_B N_B)]\}^{0.5} \tag{3.41}$$

which is the conventional space-charge-region thickness model for an abrupt heterojunction.

The barrier potentials $V_{B1}$, $V_{Bgc}$, $V_{Bgv}$, and $V_{B2}$ shown in Figure 3.18 are readily obtained from (3.37) to (3.40) as

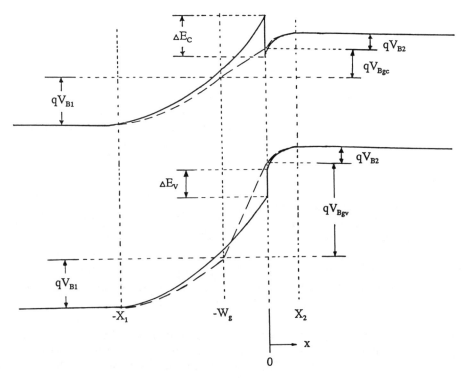

**Figure 3.18** Qualitative illustration of on the energy band diagram of the heterojunction without the graded layer (solid lines) and with the graded layer (dashed lines).

$$V_{B1} = 0.5(qN_E/\varepsilon_E)(X_1 - W_g)^2 \tag{3.42}$$

$$V_{Bgc} = -\Delta E_C/q + qN_B X_2 W_g/\varepsilon_B - 0.5qN_E W_g^2/\varepsilon_B - (qN_E/6)(0.93\varepsilon_0 W_g^2/\varepsilon_B^2) \tag{3.43}$$

$$V_{Bgv} = -\Delta E_V/q + qN_B X_2 W_g/\varepsilon_B - 0.5qN_E W_g^2/\varepsilon_B - (qN_E/6)(0.93\varepsilon_0 W_g^2/\varepsilon_B^2) \tag{3.44}$$

$$V_{B2} = 0.5(qN_B/\varepsilon_B)X_2^2 \tag{3.45}$$

For the case when $X_1 < W_g$, $V_{B1}$ approaches zero and $V_{Bgc}$ and $V_{Bgv}$ need to be modified as

$$V_{Bgc} = -\Delta E_C/q + qN_B X_2 X_1/\varepsilon_B - 0.5qN_E X_1^2/\varepsilon_B - (qN_E/6)(0.93\varepsilon_0 X_1^2/\varepsilon_B^2) \tag{3.46}$$

$$V_{Bgv} = -\Delta E_V/q + qN_B X_2 X_1/\varepsilon_B - 0.5qN_E X_1^2/\varepsilon_B - (qN_E/6)(0.93\varepsilon_0 X_1^2/\varepsilon_B^2) \tag{3.47}$$

As will be shown later, the values of $V_{Bgc}$ and $V_{Bgv}$ can be positive or negative depending on the applied voltage and thickness of the graded layer. A positive value indicates the potential has a positive slope, and vice versa if the value is negative. In fact, a positive slope implies that the spike has been effectively removed. Note that $V_{Bgc}$ and $V_{Bgv}$ reduce to $-\Delta E_C/q$ and $\Delta E_V/q$, respectively, for an abrupt junction ($W_g = 0$).

We consider a typical heterostructure with three different graded layer thicknesses $W_g = 0$, 150 Å, and 300 Å. Figures 3.19(a, b) show the barrier potentials $V_{B1}$ and $V_{B2}$ versus the applied voltage $V_{BE}$ calculated from the present model. The barrier potential $V_{B1}$ is decreased when $W_g$ is increased, but the reverse trend is found in $V_{B2}$. The barrier potentials $V_{Bgc}$ and $V_{Bgv}$ in the graded layer are given in Figures 3.20(a, b). When $W_g = 0$ (abrupt junction), both $V_{Bgc}$ ($= -0.22$V) and $V_{Bgv}$ ($= 0.15$V) are constant with respect to $V_{BE}$. Note that the negative value indicates the potential has a negative slope versus the position. It is interesting to see that for the cases of $W_g = 150$ Å and 300 Å, $V_{Bgc}$ is positive at small $V_{BE}$ and becomes negative when $V_{BE}$ is large. On the other hand, $V_{Bgv}$ are always positive and increase monotonically as $W_g$ increases.

The change in $V_{Bgc}$ and $V_{Bgv}$ versus the applied voltage can be better illustrated by the conduction- and valence-and edges ($E_C$ and $E_C$) given in Figure 3.21. In the figure, $W_g = 100$ Å and three different $V_{BE}$ are considered. Notice the change of $E_C$ in the graded layer as $V_{BE}$ increases the slope of $V_{Bgc}$ is positive when $V_{BE}$ is small and becomes negative when $V_{BE}$ increases beyond 1.1V. The negative slope of $V_{Bgc}$ in effect creates a barrier that can limit the collector current at high bias conditions. The model predications agree with numerical simulations by Tiwari and Frank [21], which suggested that an "alloy" barrier may form in the graded layer if the base-emitter bias is high and the emitter doping concentration is relatively low (less than about $5 \times 10^{17}$ cm$^{-3}$). To illustrate this further, we compare in Figure 3.22 the model predictions with simulation results

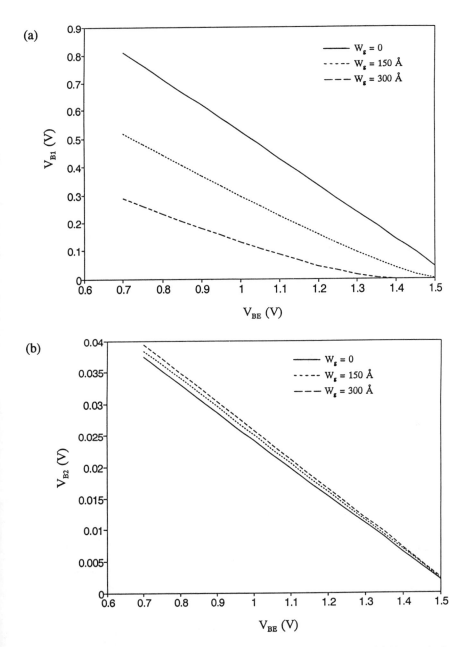

**Figure 3.19** (a) Barrier potential $V_{B1}$ on the emitter side and (b) barrier potential $V_{B2}$ on the base side calculated for three different graded layer thicknesses.

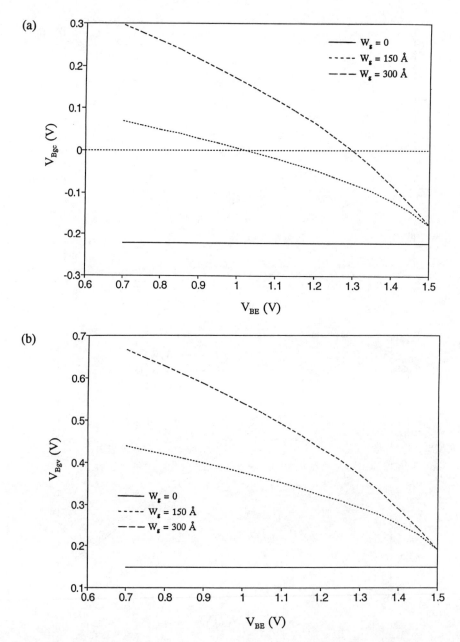

**Figure 3.20** (a) Barrier potential $V_{Bgc}$ of the conduction band in the graded layer and (b) barrier potential $V_{Bgv}$ of the valence band in the graded layer calculated for three different graded layer thicknesses.

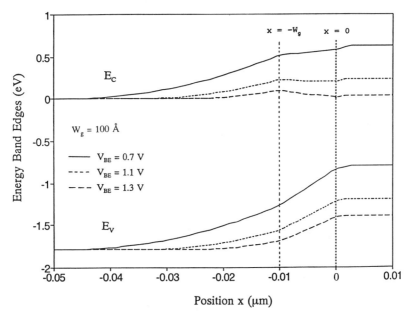

**Figure 3.21** Conduction-band and valence-band edges calculated from the present model for $W_g = 100$ Å and three different applied voltages.

in [21] for a graded HBT having $W_g = 300$ Å and a low emitter doping concentration ($N_E = 2 \times 10^{17}$ cm$^{-3}$). The presence of the alloy barrier at $V_{BE} = 1.4$V is clearly demonstrated.

### 3.3.1 Collector Current

Following the thermionic-field-diffusion approach developed by Grinberg et al. [13], the electron current density $J_n$ across the spike (located at $x = -W_g$) is the difference between two opposing fluxes

$$J_n(-W_g) = qv_n\gamma_n[n(-W_g^-) - n(-W_g^+)] \qquad (3.48)$$

Here,

$$n(-W_g^-) = N_E \exp(-V_{B1}/V_T) \quad \text{and} \quad n(-W_g^+) = n(X_2)\exp[(V_{B2} + V_{Bgc})/V_T] \qquad (3.49)$$

At this point, the only unknown parameter is $n(X_2)$, which can be solved using the relation

$$J_n(-W_g) = J_{SCRB} + J_{SCRG} + J_n(X_2) \qquad (3.50)$$

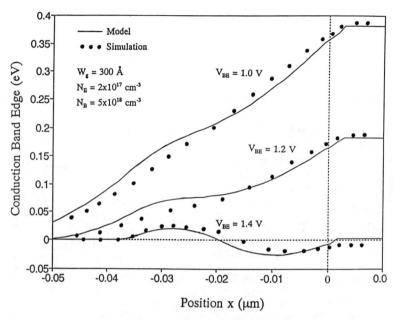

**Figure 3.22** The conduction band edges calculated from the model and obtained from simulations [21] for $W_g = 300$ Å and three different applied voltages.

where $J_{SCRB}$ is recombination current density in the space-charge layer associated with the base layer ($x = 0$ and $x = X_2$) and $J_{SCRG}$ is the recombination current density in the graded layer (models for $J_{SCRB}$ and $J_{SCRG}$ will be developed in the next section), and $J_n(X_2)$ is the diffusion-only current in the QNB. For a very thin base,

$$J_n(X_2) = J_C \approx qD_n n(X_2)/(W_B + \Delta W_B + D_n/v_s) \tag{3.51}$$

Combining the above equations and solving for $n(X_2)$, we obtain

$$n(X_2) = [qv_n\gamma_n N_E \exp(-V_{B1}/V_T) - J_{SCRB} - J_{SCRG}]/\zeta \tag{3.52}$$

where

$$\zeta = qD_n/(W_B + \Delta W_B + D_n/v_s) + qv_n\gamma_n \exp[(V_{B2} + V_{Bgc})/V_T] \tag{3.53}$$

### 3.3.2 Base Current

The $J_B$ model for the abrupt HBT (Section 3.3) is again applicable here provided (1) $\Delta E_C/q$ and $\Delta E_V/q$ are changed to $V_{Bgc}$ and $V_{Bgv}$, respectively, and (2) $J_{SCR}$ is modified to account for the electron-hole recombination in the graded layer.

$J_{SCR}$ now consists of three recombination current densities occurring in the emitter-side of the space-charge layer ($J_{SCRE}$), in the graded layer ($J_{SCRG}$), and in the base-side of the space-charge layer ($J_{SCRB}$). Thus

$$J_{SCR} = J_{SCRE} + J_{SCRG} + J_{SCRB} \tag{3.54}$$

$$= q\int_{-X_1}^{-W_g} U_{SRH,E}\,dx + q\int_{-W_g}^{0} U_{SRH,G}\,dx + q\int_{0}^{X_2} U_{SRH,B}\,dx \tag{3.55}$$

where

$$U_{SRH,E} = 0.5\sigma v_n N_t n_{iE}\exp(V_{BE}/2V_T) \tag{3.56}$$

$$U_{SRH,G} = 0.5\sigma v_n N_t n_{ig}\exp(V_{BE}/2V_T) \tag{3.57}$$

$$U_{SRH,B} = 0.5\sigma v_n N_t n_{iB}\exp(V_{BE}/2V_T) \tag{3.58}$$

Here $n_{iE}$, $n_{ig}$, and $n_{iB}$ are the intrinsic concentrations in the $Al_{0.3}Ga_{0.7}As$ emitter, graded layer, and GaAs base, respectively. Note that $n_{ig}$ is a function of the position

$$n_{ig} = n_{iE}\exp(0.5E_{GE}/kT)\exp[-0.5E_G(x)/kT] \tag{3.59}$$

where $E_{GE}$ is the energy bandgap of the $Al_{0.3}Ga_{0.7}As$ and $E_G(x) = E_{GB} - [(E_{GE} - E_{GB})/W_g]x$. If $X_1 < W_g$, then $J_{SCRE}$ becomes zero and the boundaries of $J_{SCRG}$ become $-X_1$ and 0.

We consider the same HBT make-up as that used in the previous section but without the setback layer and the three different graded layer thicknesses $W_g = 0$, 150 Å, and 300 Å. The collector current densities calculated from the present model are shown in Figure 3.23. The current increases considerably when $W_g$ is increased from 0 to 150 Å but only slightly when $W_g$ is increased from 150 Å to 300 Å. Of equal importance to note is the roll-off of $J_C$ at high $V_{BE}$ for both $W_g = 150$ Å and 300 Å, which is due to the presence of the "alloy" barrier in the graded layer discussed previously.

The change of the barrier shape in the graded layer has a significant consequence on the free-carrier tunneling mechanism in the heterojunction. To investigate this, we calculated the collector current density $J'_C$ excluding the tunneling mechanism. The comparison of $J'_C$ and $J_C$ is given in Figure 3.24. For an abrupt junction, free-carrier tunneling is important for a wide range of applied voltages. When the junction is graded, however, such a mechanism becomes less significant. Note that for $W_g = 150$ Å and 300 Å, $J'_C = J_C$ at small $V_{BE}$ because the barrier potential in the graded layer has a positive slope and the spike is effectively removed. As $V_{BE}$ is increased, however, the slope becomes negative and an "alloy" barrier is formed, thus making the tunneling mechanism more important.

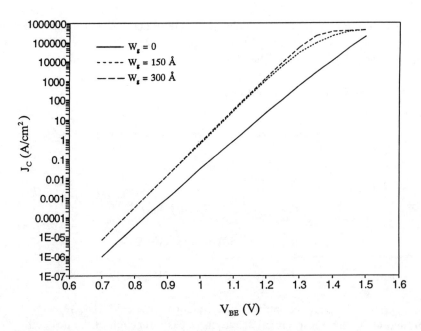

**Figure 3.23** Collector current densities versus $V_{BE}$ calculated from the present model.

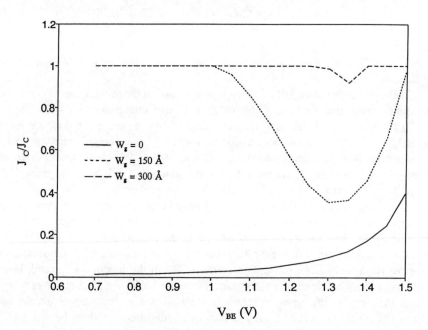

**Figure 3.24** Comparison of the collector current density without tunneling ($J'_C$) and with tunneling ($J_C$).

The graded layer also increases the base current density, as evidenced by the results shown in Figure 3.25. Again the base current increases more significantly from $W_g = 0$ to 150 Å than from 150 Å to 300 Å. The increased $J_B$ results mainly from the increase of the recombination currents in the quasi-neutral base and in the space-charge region.

Figure 3.26 shows the dc current gains β calculated from the present model for the three $W_g$. For the device make-up considered and parameters used, our calculations suggest that a graded layer of 150 Å improves β only slightly when the current level is low and more significantly when the current is high. Furthermore, it is shown that a 300-Å graded layer can actually degrade the HBT performance at low current density. Similar β are obtained for both 150-Å and 300-Å graded layers when the current density is high. These trends agree with those obtained from a more complex model derived from the charge-control concept [22].

In Figure 3.27, we compare the collector and base currents obtained from the model with experimental results of Hafizi et al. [23] for a graded HBT that has the same doping concentrations as that considered previously, a 1200-Å emitter layer thickness, a 1400-Å base layer thickness, and a 300-Å graded layer. Excellent agreement is found between the model predictions and measurements.

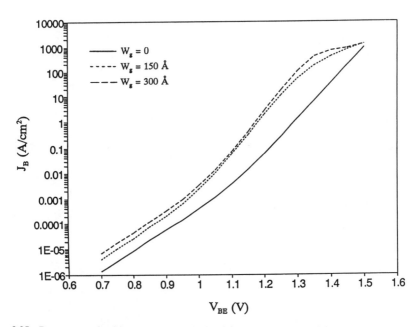

**Figure 3.25** Base current densities versus $V_{BE}$ calculated from the present model.

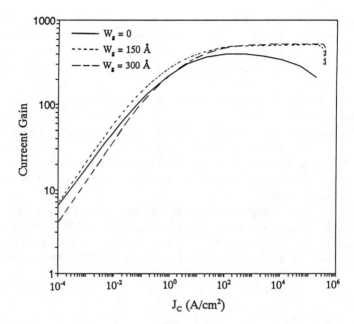

**Figure 3.26** The dc current gains versus $J_C$ calculated for three different graded layer thicknesses.

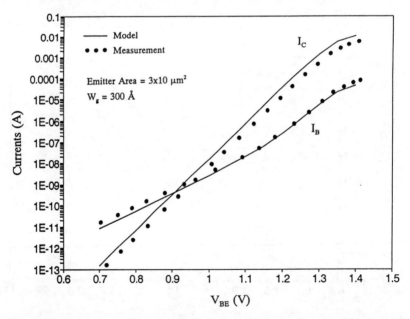

**Figure 3.27** Comparison of the collector and base currents calculated from the model and obtained from measurements (*Source*: [23]. © 1990 IEEE.).

## 3.4 COMBINED EFFECT OF SETBACK AND GRADED LAYERS

Figure 3.28 shows the device structure of an HBT with both the graded and setback layers. Using the same approach given in the previous sections, we can derive the electric fields $\xi_1(x)$ in the space-charge region on the emitter side, $\xi_s(x)$ in the setback layer, and $\xi_2(x)$ in the space-charge region on the base side:

$$\xi_1(x) = (qN_E/\varepsilon_E)(x + X_1) \quad \text{for } -X_1 < x < -W_G \tag{3.60}$$

$$\xi_g(x) = [qN_E/\varepsilon_g(x)](x + X_1) \quad \text{for } -W_g < x < 0 \tag{3.61}$$

$$\xi_s(x) = (qN_B/\varepsilon_B)(X_2 - W_I) \quad \text{for } 0 < x < W_I \tag{3.62}$$

$$\xi_2(x) = (qN_B/\varepsilon_B)(X_2 - x) \quad \text{for } W_I < x < X_2 \tag{3.63}$$

The corresponding electrostatic potential $V(x)$ can be calculated by integrating $\xi(x)$ over its boundary. Thus

$$V_1(x) = 0.5(qN_E/\varepsilon_E)(x + X_1)^2 \tag{3.64}$$

$$V_g(x) = V_1(-W_G) + qN_EW_G(x + W_G)/(\varepsilon_B - \varepsilon_E)$$

$$+ [qN_EW_G/(\varepsilon_B - \varepsilon_E)][X_1 - W_G\varepsilon_B/(\varepsilon_B - \varepsilon_E)]$$

$$\cdot \ln\{[(\varepsilon_B - \varepsilon_E)x + W_G\varepsilon_B]/(W_G\varepsilon_E)\} \tag{3.65}$$

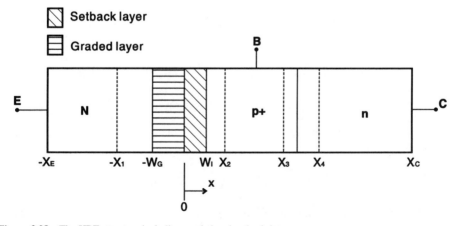

**Figure 3.28** The HBT structure including graded and setback layers.

$$V_s(x) = V_g(0) + (qN_B/\varepsilon_B)(X_2 - W_I)x \tag{3.66}$$

$$V_2(x) = V_s(W_I) + (qN_B/\varepsilon_B)(X_2 x - x^2/2 - X_2 W_I + W_I^2/2) \tag{3.67}$$

where

$$V_1(-W_G) = 0.5(qN_E/\varepsilon_E)(X_1 - W_G)^2 \tag{3.68}$$

$$V_g(0) = V_1(-W_G) + qN_E W_G^2/(\varepsilon_B - \varepsilon_E)$$
$$+ [qN_E W_G/(\varepsilon_B - \varepsilon_E)][X_1 - W_G \varepsilon_B/(\varepsilon_B - \varepsilon_E)]\ln(\varepsilon_B/\varepsilon_E) \tag{3.69}$$

$$V_s(W_I) = V_g(0) + (qN_B/\varepsilon_B)(X_2 - W_I)W_I \tag{3.70}$$

The space-charge region thickness $X_1$ can be found from (3.70) and $X_2 = (N_B/N_E)X_1 + W_I$ (derived from charge neutrality in the entire space-charge region) as

$$X_1 = -0.5B/A - 0.5(B^2 - 4AC)^{0.5}/A \tag{3.71}$$

where

$$A = 0.5qN_E/\varepsilon_E + 0.5qN_E^2/(\varepsilon_E N_B) \tag{3.72}$$

$$B = -qN_E W_G/\varepsilon_E - [qN_E W_G/(\varepsilon_B - \varepsilon_E)]\ln(\varepsilon_B/\varepsilon_E) + qN_E W_I/\varepsilon_E \tag{3.73}$$

$$C = -qN_E W_G^2/(\varepsilon_B - \varepsilon_E) + [qN_E W_G^2 \varepsilon_B/(\varepsilon_B - \varepsilon_E)^2]\ln(\varepsilon_B/\varepsilon_E) + 0.5qN_E W_G^2/\varepsilon_E \tag{3.74}$$

The barrier potentials for the conduction-band edge $E_C$ in the space-charge region are defined as

$$V_{B1} = V_1(-W_G) \qquad V_{Bgc} = -\Delta E_C/q + V_g(0) - V_1(-W_G)$$
$$V_{BS} = V_s(W_I) - V_g(0) \qquad V_{B2} = V_2(X_2) - V_s(W_I) \tag{3.75}$$

The barrier potentials for the valence band are the same as that given in (3.75) provided $V_{Bgc}$ is changed to

$$V_{Bgv} = \Delta E_v/q + V_g(0) - V_1(-W_G) \tag{3.76}$$

Once the barrier potentials are found, the collector and base currents of the HBT can be calculated using the same approach given in the previous sections.

For illustrations, we consider a typical $Al_{0.3}Ga_{0.7}As/GaAs$ HBT structure that has $5 \times 10^{17}$ cm$^{-3}$ emitter, $5 \times 10^{19}$ cm$^{-3}$ base, and $5 \times 10^{16}$ cm$^{-3}$ collector doping concentrations and 1700-Å emitter, 1000-Å base, and 5000 Å collector layer thicknesses, respectively. Also, the HBT has graded and setback layer thicknesses ranging from 0 to 300 Å. Since the results for HBTs that have only the graded layer and only the setback layer have been reported in the previous sections, our emphasis here is on the combined effects of the two layers on the HBT performance.

The voltage dependencies of the collector and base currents are plotted in Figures 3.29 and 3.30, respectively. The results suggest that, except for very large $V_{BE}$ at which the alloy junction barrier exists, both $J_C$ and $J_B$ increase considerably when $W_I$ and $W_G$ increase from 0 to 150 Å but change only slightly when $W_I$ and $W_G$ increase from 150 Å to 300 Å. The $J_C$ increase results from the removal of $\Delta E_C$, which reduces the importance of thermionic and tunneling mechanisms. On the other hand, the $J_B$ increase is due to the fact that the presence of $W_I$ and $W_G$ widens the space-charge layer and thus enhances the electron-hole recombination in the region. Again the results calculated from the present model compare favorably with those obtained from the numerical model [14]. Note that when $W_I$ and $W_G$ are present, $J_C$ saturates at a smaller $V_{BE}$ (Figure 3.29), stemming from the fact that the free-carrier transport in such HBTs is hindered by the alloy junction barrier formed at large $V_{BE}$.

To better compare the currents in HBTs with $W_I$ and $W_G$ and without $W_I$ and $W_G$ (abrupt HBT), the currents in HBTs with $W_I$ and $W_G$ are normalized by those in abrupt HBTs. These normalized $J_C$ and $J_B$ are shown in Figures 3.31 and 3.32, respectively. The results in Figure 3.31 indicate that the HBT with $W_I = W_G = 150$ Å has the highest $J_C$ at

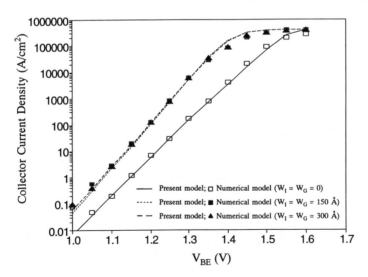

**Figure 3.29** Collector current densities calculated from the present model and the numerical model for HBTs with three different $W_I$ and $W_G$ make-ups.

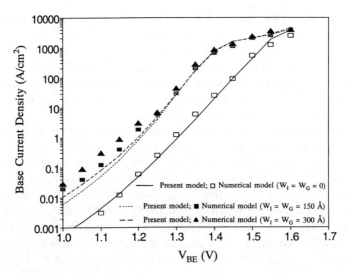

**Figure 3.30** Base current densities calculated from the present model and the numerical model for HBTs with three different $W_I$ and $W_G$ make-ups.

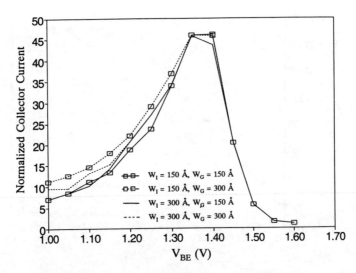

**Figure 3.31** Collector current densities of HBTs having four different $W_I$ and $W_G$ make-ups normalized by that of the abrupt HBT ($W_I = W_G = 0$).

relatively low $V_{BE}$, but all four HBTs have similar $J_C$ at high $V_{BE}$. In comparison with the abrupt HBT, $J_C$ in the HBTs with $W_I$ and $W_G$ is about 5 to 10 times larger at $V_{BE} = 1V$, increases to a maximum of about 45 times larger at $V_{BE} = 1.4V$, and becomes comparable with that in the abrupt HBT at high voltages. The presence of $W_I$ and $W_G$ also increases $J_B$, as shown in Figure 3.32. Intuitively, one would expect that $W_I = W_G = 300$ Å results in the widest space-charge layer and thus the highest $J_B$. This is indeed the case in Figure 3.32, which shows that the HBT with $W_I = W_G = 300$ Å has the highest $J_B$ at relatively small $V_{BE}$ among the four HBTs. At high $V_{BE}$, all four HBTs have similar $J_B$ because the space-charge region has been greatly reduced by the high $V_{BE}$ and the space-charge region recombination is not important. Conversely, the hole current $J_{RE}$ injected from the base to the emitter becomes the dominant component for $J_B$. Since $J_{RE}$ is not affected by $W_I$ and $W_G$ at high voltages (the valence barrier is almost flat at such bias conditions), all four HBTs have comparable $J_B$.

Figure 3.33 shows the dc current gains at a low current level ($J_C = 0.01$ A/cm$^2$ is considered) calculated as a function of $W_I$ and $W_G$. The results suggest that the current gain β decreases linearly with increasing $W_G$ if $W_I = 0$ and that b is relatively insensitive to $W_G$ if $W_I$ is greater than 150 Å. Also, at this current level, β decreases with increasing $W_I$ for all $W_G$.

The trends are quite different for the HBT operated at a high current level ($J_C = 10^5$ A/cm$^2$ is considered), however, as shown in Figure 3.34. First, comparable current gains are found for all $W_I$ and $W_G > 150$ Å. Furthermore, when $W_G$ approaches zero, having a nonzero $W_I$ is beneficial, but increasing $W_I$ beyond 150 Å does not provide any current gain enhancement.

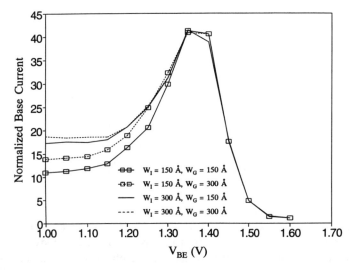

**Figure 3.32** Base current densities of HBTs having four different $W_I$ and $W_G$ make-ups normalized by that of the abrupt HBT ($W_I = W_G = 0$).

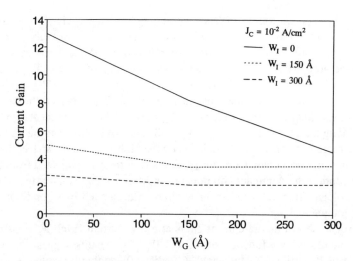

**Figure 3.33** HBT current gains at a low current level calculated as a function of $W_G$ and $W_I$.

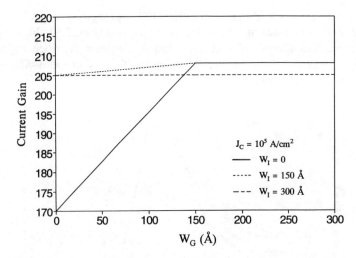

**Figure 3.34** HBT current gains at a high current level calculated as a function of $W_G$ and $W_I$.

The current gain and cutoff frequency of the HBTs that have an abrupt junction, abrupt junction with a setback layer (150 Å), graded layer (150 Å), and setback and graded layers are summarized in Figures 3.35(a, b), respectively. The trends suggest that the optimal peak current gain and cutoff frequency can be obtained by using both the setback and graded layers. It is important to point out that the current gain and cutoff frequency shown in Figures 3.35(a, b) do not exhibit a rapid fall-off at high current levels—a

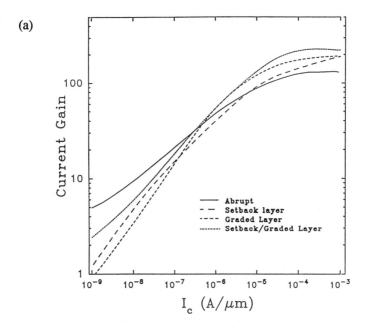

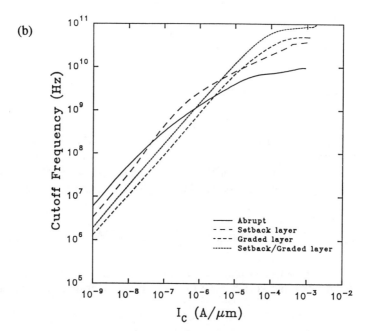

**Figure 3.35** Comparison of the (a) current gain and (b) cutoff frequency of HBTs having an abrupt junction, an abrupt junction with a setback layer (150 Å), a graded layer (150 Å), and setback and graded layers.

phenomenon commonly observed in measurements. This is due to the fact that an important mechanism in the AlGaAs/GaAs HBT, called the self-heating effect, has been neglected in the models developed. Such an effect will be treated in Chapter 4.

## 3.5 PASSIVATION EMITTER LEDGE

Typical emitter-up AlGaAs/GaAs HBTs are fabricated with mesa etching techniques that create exposed extrinsic base surface and emitter side-walls, as indicated by the bold lines in Figure 3.36(a). Because free GaAs and AlGaAs surfaces are characterized by a high surface recombination velocity, recombination current taking place at these surfaces can be a major component to the overall base current, particularly if the emitter area is small. To increase the current gain, passivation to these areas by a thin depleted AlGaAs layer (ledge) is often used [24]. As can be seen from Figure 3.36(b), the thickness of the ledge layer is the same as that of the AlGaAs emitter, but the passivation ledge distance (between the emitter and base contacts) should be long enough (e.g., 3000 Å) to suppress the base contact recombination [25]. Figure 3.37 shows the measured current gains as a function of the collector current for the passivated and unpassivated HBTs with an emitter area of $4 \times 10$ μm$^2$ [5]. Clearly, the current gain of the HBT is improved considerably if a passivation ledge structure is incorporated. The HBT current gain is also a strong function of the ledge length and the undepleted ledge thickness $\Lambda$, as demonstrated by a two-dimensional analysis [25]. Figure 3.38 shows the current gain simulated from such an analysis. In general, the current gain increases with increasing ledge length, and the highest current gain is obtained when the ledge is fully depleted (e.g., $\Lambda = 0$).

## 3.6 PROTON-IMPLANTED COLLECTOR

High cutoff frequency $f_T$ is one of the main attractions of the HBT. To further improve $f_T$ in HBTs, a process has also been used in which protons (O$^+$ or H$^+$) are implanted in the extrinsic collector region (Figure 3.39) to reduce the total base-collector junction capacitance $C_{BC}$ [26]. Let us consider an N/p$^+$/n/n$^+$ AlGaAs/GaAs/GaAs HBT with proton-implanted collector (Figure 3.39). As discussed in Chapter 2, the cutoff frequency is given by

$$f_T = 1/[2\pi(\tau_E + \tau_C + \tau'_{CT} + \tau_{BT})] \tag{3.77}$$

Here we focus on the collector capacitance charging time $\tau_C$ because it is directly proportional to $C_{BC}$. It can be expressed in terms of intrinsic and extrinsic collector make-ups as

$$\tau_C = r_{C.IN}C_{BC.IN}A_{BC.IN} + r_{C.EX}C_{BC.EX}A_{BC.EX} \tag{3.78}$$

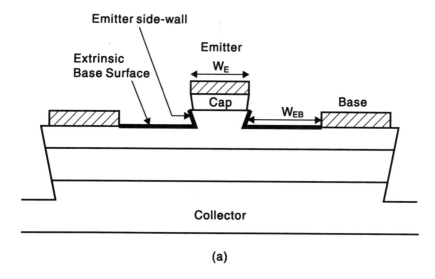

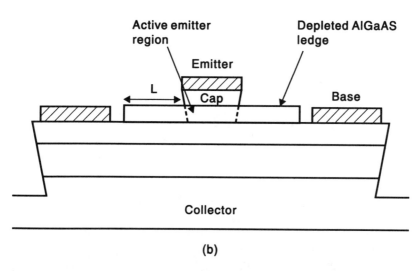

**Figure 3.36** (a) An HBT without a passivation ledge structure, with emitter side-walls and extrinsic base surface emphasized, and (b) an HBT with an AlGaAs passivation ledge.

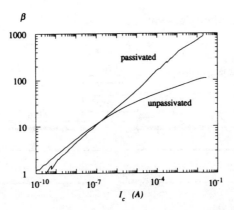

**Figure 3.37** Current gains measured from HBTs with and without passivation ledge (*Source*: [5]. © 1992 IEEE.).

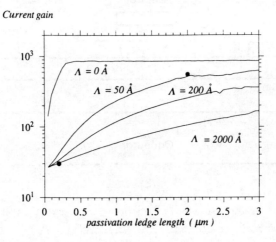

**Figure 3.38** Simulated current gains as a function of passivation ledge length and undepleted ledge thickness $\Lambda$. The two closed circles are measured data [25].

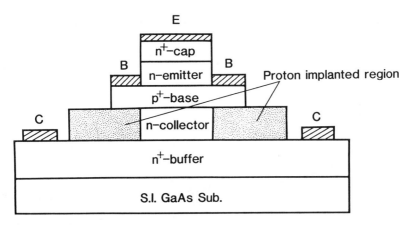

**Figure 3.39** HBT structure with proton-implanted collector.

where subscripts IN and EX denote intrinsic and extrinsic, respectively; $r_C$ is the collector resistance; and $A_{BC}$ is the base-collector junction area. $C_{BC.N}$ can be approximated by the conventional depletion capacitance model, which is proportional to $V_{CB}^{0.5}$. On the other hand, $C_{BC.EX}$ is bias-independent:

$$C_{BC.EX} = A_{BC.EX} \varepsilon_C / W_C \qquad (3.79)$$

where $\varepsilon_C$ is the collector dielectric permittivity and $W_C$ is the collector thickness. Note that in the conventional treatment, which is invalid for the HBT with proton-implanted collector,

$$\tau_C = r_{C.IN} C_{BC.IN} (A_{BC.IN} + A_{BC.EX}) \qquad (3.80)$$

For illustrations, we consider an HBT that has an emitter-base junction area of 20 μm², base-collector junction area of 80 μm² (e.g., extrinsic base-collector junction area of 60 μm²), $r_{C.IN} = 20\ \Omega$, and $r_{C.EX} = 400\ \Omega$ [6]. Figure 3.40 compares values for $\tau_C$ calculated from the above model (refer to (3.78)) and from the conventional model (see (3.80)). It is shown that $\tau_C$ in such an HBT depends less strongly on the applied collector-base voltage $V_{CB}$ than that in the HBT without a proton-implanted collector due to the fact that the bias-independent $C_{BC.EX}$ dominates the total base-collector junction capacitance of the proton-implanted HBT. Furthermore, at relatively small $V_{CB}$, a smaller $\tau_C$ is found when the proton-implanted collector is considered. As evidenced in Figure 3.41, this leads to a higher $f_T$ for relatively small $V_{CB}$ (or $V_{CE}$) (e.g., $V_{CE}$ between 1.5V to 5.0V), and the conventional model fails to give an accurate description of the $f_T$ behavior in this bias region. The proton-implanted collector appears to offer no enhancement on $f_T$, however, as $V_{CE}$ is increased beyond 5V.

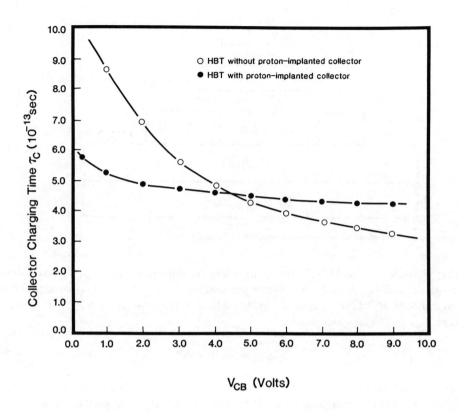

**Figure 3.40** Collector charging time of an HBT calculated with and without proton-implanted collector.

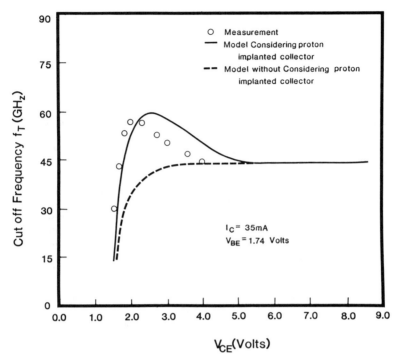

**Figure 3.41** Comparison of cutoff frequencies calculated from the model without considering proton-implanted collector, calculated from the model considering proton-implanted collector, and obtained from measurements [6].

# References

[1] Ito, H., T. Ishibashi, and T. Sugeta, "Current Gain Enhancement in Graded Base AlGaAs/GaAs HBTs Associated with Electron Drift Motion," *Jap. J. Appl. Phys.,* Vol. 24, April 1985, pp. L241–L243.
[2] Miller, D. L., P. M. Asbeck, R. J. Anderson, and F. H. Eisen, "(GaAl)As/GaAs Heterojunction Bipolar Transistors with Graded Composition in the Base," *Electron. Lett.,* Vol. 19, May 1983, pp. 367–368.
[3] Hayes, J. R., F. Capasso, A. C. Gossard, R. J. Malik, and W. Wiegmann, "Bipolar Transistor With Graded Band Gap Base," *Electron. Lett.,* Vol. 19, May 1983, pp. 411–412.
[4] Liou, J. J., W. W. Wong, and J. S. Yuan, "A Study of the Base Built-in Field on the Steady-State Current Gain of Heterojunction Bipolar Transistors," *Solid-St. Electron.,* Vol. 33, 1990, pp. 845–849.
[5] Liu, W., and J. S. Harris, Jr., "Diode Ideality Factor for Surface Recombination Current in AlGaAs/GaAs Heterojunction Bipolar Transistors," *IEEE Trans. Electron Devices,* Vol. 39, 1992, p. 2726.
[6] Inada, M., Y. Ota, A. Nakagawa, M. Yanagihara, T. Hirose, and K. Eda, "AlGaAs/GaAs Heterojunction Bipolar Transistors with Small Size Fabricated by a Multiple Self-Alignment Processes Using One Mask," *IEEE Trans. Electron Devices,* Vol. ED-34, 1987, p. 2405.
[7] Rockett, P. I., "Monte Carlo Study of the Influence of Collector Region Velocity Overshoot on the High-Frequency Performance of AlGaAs/GaAs Heterojunction Bipolar Transistors," *IEEE Trans. Electron Devices,* Vol. ED-35, October 1988, pp. 1573–1579.

[8] Horio, K., Y. Iwatsu, and H. Yanai, "Numerical Simulation of AlGaAs/GaAs Heterojunction Bipolar Transistors with Various Collector Parameters," *IEEE Trans. Electron Devices*, Vol. ED-36, April 1989, pp. 617–624.

[9] Katoh, R., M. Kurata, and J. Yoshida, "Self-Consistent Particle Simulation for (AlGa)As/GaAs HBT's with Improved Base-Collector Structures," *IEEE Trans. Electron Devices*, Vol. ED-36, May 1989, pp. 846–853.

[10] Klausmeier-Brown, M. E., M. S. Lundstrom, and M. R. Melloch, "The Effects of Heavy Impurity Doping on AlGaAs/GaAs Bipolar Transistors," *IEEE Trans. Electron Devices*, Vol. 36, May 1989, pp. 2146–2155.

[11] Liou, J. J., "Effects of Base Grading on the Early Voltage of HBT's," *Solid-St. Electron.*, 1992, Vol. 35, p. 1378.

[12] Liou, J. J., C. S. Ho, L. L. Liou, and C. I. Huang, "An Analytical Model for Current Transport in AlGaAs/GaAs Abrupt HBTs with a Setback Layer," *Solid-St. Electron.*, Vol. 36, 1993, pp. 819–825.

[13] Grinberg, A. A., M. S. Shur, R. J. Fischer, and H. Morkoc, "An Investigation of the Effect of Graded Layers and Tunneling on the Performance of AlGaAs/GaAs Heterojunction Bipolar Transistors," *IEEE Trans. Electron Devices*, Vol. ED-31, 1984, p. 1758.

[14] Liou, L. L., and C. I. Huang, "Using Constant Base Current as a Boundary Condition for One-Dimensional AlGaAs/GaAs Heterojunction Bipolar Transistor Simulation," *Electron Lett.*, Vol. 26, 1990, p. 1501.

[15] Chen, S.-C., Y.-K. Su, and C.-Z. Lee, "A Study of Current Transport on p-N Heterojunctions," *Solid-St. Electron.*, Vol. 35, 1992, p. 1311.

[16] Kroemer, H., "Heterostructure Bipolar Transistor and Integrated Circuits," *Proc. IEEE*, Vol. 70, 1982, p. 13.

[17] Marty, A., G. E. Rey, and J. P. Bailbe, "Electrical Behavior of an Npn GaAl/GaAs Heterojunction Transistor," *Solid-St. Electron.*, Vol. 22, 1979, p. 549.

[18] Grinberg, A. A., and S. Luryi, "On the Thermionic-Diffusion Theory of Minority Transport in Heterostructure Bipolar Transistors," *IEEE Trans. Electron Devices*, Vol. 40, 1993, p. 859.

[19] Adachi, S., "GaAs, AlAs, and $Al_xGa_{1-x}As$: Material Parameters for Use in Research and Device Applications," *J. Appl. Phys.*, Vol. 58, August 1985, pp. R1–R29.

[20] Liou, J. J., "An Analytical Model for the Current Transport in Graded Heterojunction Bipolar Transistors," *Solid-St. Electron.*, Vol. 38, 1995, p. 946.

[21] Tiwari, S., and D. J. Frank, "Analysis of the Operation of GaAlAs/GaAs HBT's," *IEEE Trans. Electron Devices*, Vol. 36, 1989, pp. 2105–2121.

[22] Parikh, C. D., and F. A. Lindholm, "Space-Charge Region Recombination in Heterojunction Bipolar Transistors," *IEEE Trans. Electron Devices*, Vol. 39, 1992, p. 2197.

[23] Hafizi, M. E., C. R. Crowell, and M. E. Grupen, "The dc Characteristics of GaAs/AlGaAs Heterojunction Bipolar Transistors Applications to Device Modeling," *IEEE Trans. Electron Devices*, Vol. 37, 1990, p. 2121.

[24] Hayama, N., and K. Honjo, "Emitter Size Effect on Current Gain in Fully Aligned AlGaAs/GaAs HBTs with AlGaAs Surface Passivation Layer," *IEEE Electron Device Lett.*, Vol. 11, 1990, p. 388.

[25] Liu, W., and J. S. Harris, Jr., "Parasitic Conduction Current in the Passivation Ledge of AlGaAs/GaAs Heterojunction Bipolar Transistors," *Solid-St. Electron.*, Vol. 35, 1992, p. 891.

[26] Liou, J. J., "Modeling the Cutoff Frequency of AlGaAs/GaAs Heterojunction Bipolar Transistors With Proton-Implanted Collector Region," *Solid-St. Electron.*, Vol. 33, 1990, p. 1329.

# Chapter 4
# Thermal Effect in an AlGaAs/GaAs HBT

The advance of AlGaAs/GaAs heterojunction bipolar transistor (HBT) technology in recent years has made high output power possible and practical. The HBT's very high current-handling capability and the very poor thermal conductivity of GaAs, however, often lead to a significant increase in the lattice temperature of the HBT. This mechanism, called the self-heating effect, consequently confines the HBT performance considerably below its electronic limitation [1]. The self-heating effect is the sole thermal mechanism in an HBT that has a single emitter structure (single-finger HBT).

For modern microwave HBTs, a multiple emitter finger structure has frequently been used (multifinger HBT) in which several HBT emitters, each with its own HBT operation, are arranged in parallel to each other with proper spacing [2]. Such a structure allows less current to be carried and thus less heat power to be generated in each HBT unit cell, thus making the self-heating effect less prominent compared to its single-emitter finger counterpart. Recently, a 12.5-W cw (power density of 1.74 mW/$\mu m^2$, where $\mu m^2$ is the total emitter area) monolithic amplifier constructed using 12-finger HBTs was demonstrated at 10 GHz [3].

The thermal effect in the multifinger HBT is more complicated than that in the single-finger HBT, because thermal-coupling among the neighboring fingers (thermal-coupling effect) is also important in the multifinger HBT. The combination of self-heating and thermal-coupling effects often results in a thermally limited phenomenon called thermal runaway [4]. When the base current is fixed and is relatively large, the collector current decreases sharply (i.e., called thermal runaway) as the collector-emitter voltage is increased beyond a critical value. A common remedy to this problem is to use large resistors (ballast resistors) at the emitter and/or base contacts [5]. Such a resistance gives rise to a large voltage drop at the contact and thus reduces the voltage drop across the emitter-base junction. This in turn decreases the current density and the heat generated in each unit HBT. Evidently, this approach limits the output power density, not by thermal, but by electronic means. It can also degrade the HBT high-frequency performance due to the extra delay time through the ballast resistance.

## 4.1 SELF-HEATING EFFECT IN SINGLE-EMITTER FINGER HBTs

Consider the typical N/p/n $Al_{0.3}Ga_{0.7}As/GaAs$ single-emitter HBT shown in Figure 4.1. The following assumptions are employed in the analysis:

1. Boltzmann statistics are assumed throughout.
2. Two- or three-dimensional effects are not considered.
3. Since the size of the intrinsic HBT (e.g., intrinsic emitter, base, and collector) is much smaller than that of the extrinsic HBT (e.g., subcollector and *semi-insulating* (SI) substrate), we assume that the temperature in the intrinsic HBT, which can be much higher than the room temperature, is spatially independent and that the heat generated in the intrinsic HBT is dissipated primarily through an effective heat diffusion area in the SI substrate (Figure 4.1).

### 4.1.1 Equilibrium Free-Carrier Concentration at High Temperatures

For the n-type emitter, the thermal equilibrium majority and minority free-carrier concentrations ($n_{0,E}$ and $p_{0,E}$) at any lattice temperature $T$ are

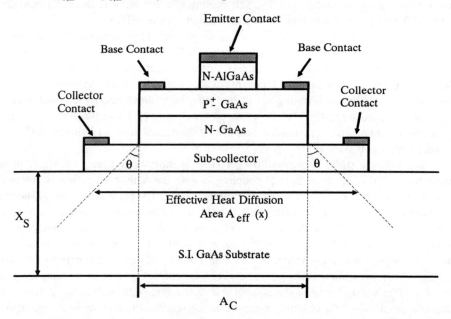

**Figure 4.1** HBT device structure illustrating the effective area through which the heat generated in the intrinsic HBT is dissipated.

$$n_{0,E} = n'_{iE} + N_E + p_{0,E} - n'_{iE} \qquad n_{0,E}p_{0,E} = n_{iE}^2 \qquad (4.1)$$

where $n'_{iE}$ is the number of free carriers generated from the valence to conduction bands [6], $n_{iE}$ is the intrinsic free-carrier concentration in the emitter ($n_{iE}$ depends strongly on the lattice temperature [6]), and $N_E^+$ is the concentration of impurity ions in the emitter resulting from the ionization process. If $T$ is in the vicinity of room temperature, $N_E^+$ equals the emitter doping concentration $N_E$, $n'_{iE}$ is not significant, and $n_{O,E} = N_E$ and $p_{O,E} = n_{iE}^2/N_E$. For higher temperatures, however, $n'_{iE}$ is important, and $n_{O,E}$ and $p_{O,E}$ need to be calculated from (4.1).

Similarly,

$$n_{0,B}p_{0,B} = n_{iB}^2 = (n'_{iB} + N_B)n'_{iB} \qquad (4.2)$$

for the p-type base and

$$n_{0,C}p_{0,C} = n_{iC}^2 = (n'_{iC} + N_C)n'_{iC} \qquad (4.3)$$

for the n-type collector. In the above equations, subscripts B and C denote the base and collector, respectively. The temperature-dependent energy bandgaps for the AlGaAs emitter and GaAs base and collector, which are needed to calculate $n_{iE}$, $n_{iB}$, and $n_{iC}$ [6], can be expressed as [7]

$$E_{GE} = 1.87 - 5.4 \times 10^{-4} T^2/(T+204) \qquad (4.4)$$

$$E_{GB} = E_{GC} = 1.52 - 5.4 \times 10^{-4} T^2/(T+204) \qquad (4.5)$$

where $E_{GE}$, $E_{GB}$, and $E_{GC}$ are the emitter ($Al_{0.3}Ga_{0.7}As$), base (GaAs), and collector (GaAs) energy bandgaps, respectively.

### 4.1.2 Lattice Temperature and Collector Current Density

Here we model the interacting behavior of collector current density $J_C$ and lattice temperature of the HBT. To simplify the analysis, a graded HBT is considered, and the drift-diffusion theory is used to describe free-carrier transport across the graded emitter-base heterojunction. This is a good assumption provided the graded layer is relatively thick (> 100 Å), as suggested by the numerical results reported in [8] and Chapter 3. Thus

$$J_C = qD_n \Delta n(0)/(W_B + \Delta W_B + D_n/v_s) \qquad (4.6)$$

where $D_n$ is the electron diffusion coefficient in the base, $W_B$ is the base layer thickness, $\Delta W_B$ is the current-induced base pushout, and $\Delta n(0)$ is the excess minority free-carrier concentration at the emitter junction. The temperature-dependent $Dn$ and $\Delta_n$ are given by [9]

$$D_n = V_T[7200/(1 + 5.5 \times 10^{-17}N_B)^{0.233}](300/T)^{2.3} \tag{4.7}$$

$$\Delta n(0) = n_{0,E}\exp[-(V_{bi,BE} - V_{j,BE})V_T] \tag{4.8}$$

Here $V_T = kT/q = 0.026(T/300)$ is the thermal voltage, $V_{j,BE}$ is the emitter-base junction voltage, and $V_{bi,BE}$ is the base-emitter junction built-in potential. Accounting for the voltage drops in the emitter and base layers, $V_{j,BE}$ is related to the applied emitter-base voltage $V_{BE}$ as

$$V_{j,BE} = V_{BE} - r_E J_C - r_B J_B \tag{4.9}$$

where $r_E$ and $r_B$ are the emitter and base series resistances ($\Omega/\text{cm}^2$), respectively, and $J_B$ is the base current density. Note that for the case of $V_{j,BE} > V_{bi,BE}$, $V_{bi,BE} - V_{j,BE}$ is set to zero to ensure that the emitter cannot unphysically inject more free carriers into the base than its impurity doping concentration. Also note that high-level injection in the base, an effect that occurs frequently in silicon *bipolar transistors* (BJTs), is not likely to happen in HBTs because the base is doped more heavily than the emitter.

The heat power $P_s$ ($W$) generated in the HBT is

$$P_s = J_C V_{CE} A_E \tag{4.10}$$

where $A_E$ is the emitter area and $V_{CE} = V_{BE} + V_{CB}$ is the applied collector-emitter voltage. According to assumption 3, the heat generated in the HBT is primarily dissipated through the SI GaAs substrate. Thus $P_s$ is related to the thermal resistance $R_{th}$ of the substrate as

$$T - T_0 = P_s R_{th} \tag{4.11}$$

where $T_0 = 300\text{K}$ is the ambient temperature. Note that the heat is dissipated throughout the SI substrate with a lateral diffusion angle $\theta$ (Figure 4.1) and that thermal conductivity $K_s$ of GaAs is proportional to $(T/T_0)^{-b}$, where $b = 1.22$ [10]. This, together with the Kirchhoff transformation [11], yields

$$R_{th} = (\eta - T_0)/P_s \tag{4.12}$$

where

$$\eta = [1/T_0^{b-1} - (b-1)R_{th0}P_s/T_0^b]^{-1/b-1} \tag{4.13}$$

Here $R_{th0}$ is the thermal resistance for the case that the thermal conductivity in the GaAs substrate is temperature independent and is denoted

$$R_{th0} = 1/K_{s0} \int_0^{X_s} dx/A_{eff}(x) = 1/K_{s0} \int_0^{X_s} dx/[A_C + 2Z\tan(\theta)x] \quad (4.14)$$

where $K_{s0}$ is the thermal conductivity at $T_0$ ($K_{s0}$ = 0.47 W/K-cm at 300K), Z is the HBT width, $A_C$ is the collector area, and $X_s$ is the thickness of the SI substrate. An angle of $\theta$ = 45 deg, which has been suggested in [12] for cases where there are no severe thermal conductivity mismatches at the thermal boundary, is used in our calculations.

The initial value of T can be calculated from (4.10) to (4.14) after the initial $J_C$ is calculated under room temperature. The correct T and $J_C$ are obtained after several iterations.

An N/p$^+$/n AlGaAs/GaAs HBT with the configuration listed in Table 4.1 under forward-active operation is used in calculations. Figure. 4.2 plots the collector and base current densities versus the base-emitter applied voltage calculated from the present model, which includes the thermal effect (hereafter called present model), and the model that assumes a constant lattice temperature of 300K (hereafter called isothermal model). Also shown are the results obtained from the numerical model reported in [13] that solves the Poisson and continuity equations coupled with the heat transfer mechanism and takes into account the nonuniform spatial temperature and band distribution and carrier degeneracy. Good agreement is found between the present model and numerical simulation. The isothermal, present, and numerical models predict the same current densities when the applied voltage is relatively small, but large discrepancies arise when the collector current density is high because the high current density generates a large amount of heat in the HBT, which results in a much higher lattice temperature in the HBT than the ambient temperature. Note that when the thermal effect is included, the collector and base current

**Table 4.1**
HBT Device Structure Used in Calculations

|  | Thickness (Å) | Type | Doping Density (/cm³) | AlAs Fraction in $Al_xGa_{1-x}As$ |
|---|---|---|---|---|
| Emitter | 1700 | n | $5 \times 10^{17}$ | 0.3 |
| Base | 1000 | p$^+$ | $1 \times 10^{19}$ | 0 |
| Collector | 3000 | n | $5 \times 10^{16}$ | 0 |

Emitter Area = 1 × 10 mm²
Collector Area = 5 × 20 mm²
S. I. Substrate Thickness = 100 mm
Lateral Heat Diffusion Angle (q) = 45 degrees
Emitter Contact Resistance = $10^{-6}$ Ω/cm²
Base Contact Resistance = $10^{-6}$ Ω/cm²
Collector Contact Resistance = $10^{-6}$ Ω/cm²

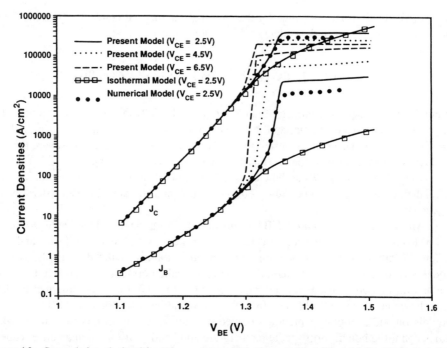

**Figure 4.2** Gummel plot calculated from the present model for $V_{CE}$ = 2.5V, 4.5V, and 6.5V. Also included are the results calculated from the isothermal model and the numerical model for $V_{CE}$ = 2.5V.

densities increase faster and become constant at a smaller $V_{BE}$ as $V_{CE}$ is increased (Figure 4.2) because the thermal effect reduces the electron mobility and the critical current at which the Kirk effect takes place. The base and collector current saturation at high $V_{BE}$ is also caused by the emitter and base series resistances. In addition, as discussed in the Chapter 3, an "alloy" barrier may form in the graded layer if the base-emitter bias is high and the emitter doping concentration is relatively low (less than about $5 \times 10^{17}$ cm$^{-3}$). Such a barrier can further limit the base and collector currents.

The relation between the HBT lattice temperature and the collector current density is given in Figure 4.3. The lattice temperatures of the HBT increase rapidly as $J_C$ is increased beyond $10^4$ A/cm$^2$ and reach 550K, 650K, and 750K at the maximum current for $V_{CE}$ = 2.5V, 4.5V, and 6.5V, respectively.

Figure 4.4 shows that the forward-active $J_C$-$V_{CE}$ characteristics calculated from the present model compare favorably with those obtained from the numerical simulations. The I-V characteristics exhibit a negative slope when the base current density is high (or $V_{BE}$ is large), which is a phenomenon caused by the self-heating effect and commonly observed in AlGaAs/GaAs HBT dc measurements.

The current gain and cutoff frequency versus $J_C$ calculated from the present model, isothermal model, and numerical model are compared in Figures 4.5 and 4.6, respectively.

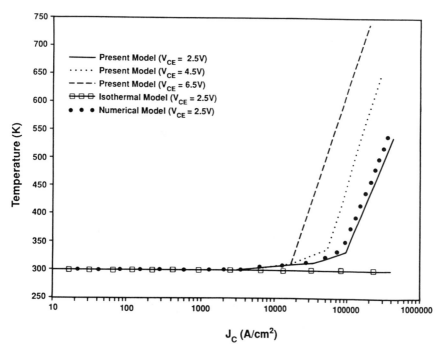

**Figure 4.3** The lattice temperature in the HBT plotted as a function of the collector current density.

It is interesting to see that both $\beta$ and $f_T$ are reduced monotonically by the increase in $V_{CE}$ at high current levels. This results because the heat power generated in the HBT is directly proportional to $V_{CE}$; hence, the thermal effect becomes more prominent at higher $V_{CE}$.

The above results are calculated using an emitter contact resistivity of $10^{-6}$ $\Omega/cm^2$. It is important to point out that the value of such resistivity depends on the emitter doping concentration as well as the Al mole fraction [14], and increasing the emitter resistance (called ballast resistance) will give rise to a smaller junction voltage, thus reducing the collector current and making the self-heating effect less prominent [15].

The results clearly show that the commonly observed rapid falloff behavior of current gain and cutoff frequency at high current density can be accurately predicted by the present model developed based on the simple drift-diffusion theory including the thermal effect. The rapid falloff, which degrades the HBT performance, is caused mainly by the high temperature associated with the high collector current and large collector-emitter applied voltage. The slower falloff predicted by the isothermal model results from the voltage drops in the emitter and base quasi-neutral regions and current-induced base pushout occurring at high current densities. For silicon BJTs the thermal effect is less important because silicon has a larger thermal conductivity than GaAs. The high-level injection in the base becomes the main contributor to the falloff characteristics of silicon

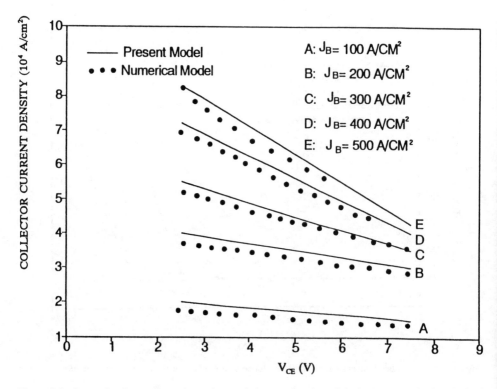

**Figure 4.4** Forward-active current-voltage characteristics as a function of the base current density obtained from the present model and the numerical model.

BJTs. Such an effect, however, is not likely to occur in HBTs because of the high base doping concentration.

## 4.2 SELF-HEATING AND THERMAL-COUPLING EFFECTS IN MULTIEMITTER FINGER HBTs

Despite the fact that the multifinger HBT has become increasingly important and popular in high-power microwave applications, efforts to analytically model such a device have been limited in the past due in part to the complicated nature of the negative feedback of the self-heating effect on the HBT current-voltage characteristics. The problem is further compounded by the thermal coupling between the neighboring emitter fingers when a multifinger structure is considered. A few numerical models have been reported in the literature [4, 15–17]. For example, solving the three-dimensional heat transfer equation, Gao et al. [15] studied the temperature in each emitter finger as a function of the emitter spacing, the number of the fingers, and the geometry of the substrate. Their results,

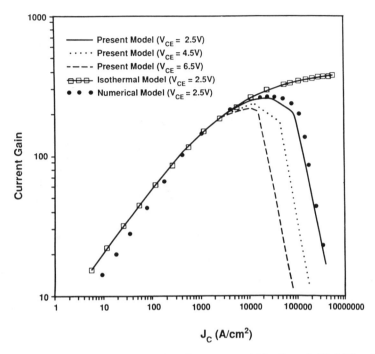

**Figure 4.5** Current gains versus $J_C$ calculated from the present model for $V_{CE}$ = 2.5V, 4.5V, and 6.5V. Also included are the results calculated from the isothermal model and the numerical model for $V_{CE}$ = 2.5V.

however, are not directly suited for HBT design because the heat power on each finger, which is related to the current and applied voltage, was treated as an independent input parameter. A self-consistent two-dimensional model was also developed by Marty et al. [17] to investigate the coupled electrothermal problem in AlGaAs/GaAs HBTs. In this approach, the current and temperature variations in the HBT are computed by considering a series of cells within the device, and the two-dimensional heat transfer in the HBT is represented by an electrical circuit analog involving vertical and horizontal thermal resistances. Such an equivalent circuit is then incorporated into a circuit simulator to simulate the HBT performance. The model, however, is limited to single-emitter finger HBTs and requires the use of a circuit simulator. Recently, an analytical HBT model including the self-heating effect was derived from the knowledge of HBT make-up [18], but it again is only applicable for the single-finger HBT.

In this section, an analytical model capable of predicting the multifinger HBT current-voltage characteristics, including the self-heating and thermal-coupling effects, will be developed. The approach given in the previous section can be extended to modeling the multifinger HBT in which several emitter fingers are arranged in parallel with proper spacing, as shown in Figure 4.7. A numerical analysis [15] solving the coupled

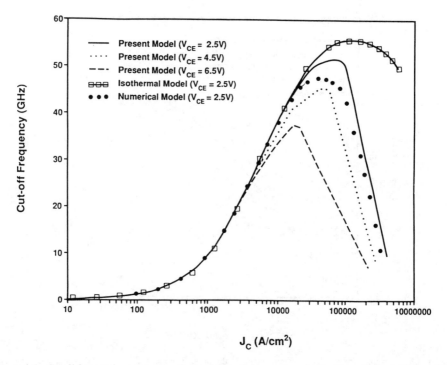

**Figure 4.6** Cutoff frequencies versus $J_C$ calculated from the present model for $V_{CE}$ = 2.5V, 4.5V, and 6.5V. Also included are the results calculated from the isothermal model and the numerical model for $V_{CE}$ = 2.5V.

current and heat transfer equations for the multifinger HBT indicates that the device performance is affected strongly by both the self-induced thermal resistance $R_{th}$ (discussed in the previous section) and the coupled thermal resistance $R_c$ due to the heating from the neighboring emitter elements. Since the thermal coupling on the subject finger due to the nearest (primary) finger is much larger than that due to the secondary fingers, we will neglect the secondary thermal coupling effect.

To illustrate the modeling concept, let us consider a 3-finger pattern. Due to the symmetrical geometry, the two outer fingers will have identical thermal properties. The temperatures at the outer and center fingers are

$$T_S = T_0 + P_{s,S} R_{th,S} + P_{s,C} R_{c,C} \tag{4.15}$$

$$T_C = T_0 + P_{s,C} R_{th,C} + 2 P_{s,S} R_{c,S} \tag{4.16}$$

where the subscripts S and C denote outer and center fingers, respectively. The term involving $R_{th}$ is the temperature rise due to self-heating in the unit HBT, and the term

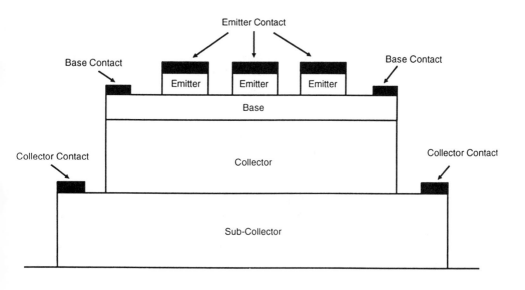

**Figure 4.7** Schematic of the multiemitter finger HBT (three emitter fingers are shown).

involving $R_c$ is the temperature rise due to thermal coupling between the subject finger and the nearest (primary) neighboring finger(s). Note that the center finger is subjected to two thermal couplings whereas the outer fingers are subjected to only one thermal coupling.

The value of $R_c$ depends on the geometry of emitter fingers and the process, including the emitter mesa etching and metalization, and is too complicated to model physically. For typical mesa-etch HBTs, $R_c$ has been empirically determined as

$$R_c = \eta R_{th}(10/S)^{1.5} \qquad (4.17)$$

where $\eta$ is a fitting parameter (i.e., for the HBT under study, using $\eta = 0.25$ gives a good agreement between the model and measurements) and $S$ is the emitter-finger spacing (in µm). Decreasing $S$ will increase $R_c$ and subsequently increase the likelihood of thermal runaway.

The initial value of $T_S$ and $T_C$ can be calculated using the above equations and model developed in Section 4.1 after the initial $J_C$, $J_B$, and $P_s$ are calculated under room temperature. These temperatures are then used to calculate the initial $R_{th}$ and $R_c$ for the three emitter fingers. The correct $T_S$, $T_C$, $J_C$, and $J_B$ for the individual emitter finger are obtained after several iterations. Summing $J_C$ and $J_B$ in each finger then yields the total $J_C$ and $J_B$ for the multifinger HBT.

We first illustrate the temperature distributions in a 3-finger HBT simulated from a numerical model [4] solving the two-dimensional heat transfer equations. Each finger has an area of $10 \times 2$ µm$^2$, and the finger spacing is 10 mm. Figures 4.8(a-d) show the two-dimensional contours of the temperature distribution at the surface of such a device biased at a constant base current of 2 mA and $V_{CE}$ of 2V, 3V, 5V, and 6V, respectively. At relatively small $V_{CE}$, all three fingers have similar temperatures, with the center finger slightly hotter than the outer fingers (Figures 4.8(a, b)). At higher $V_{CE}$, however, the thermal-coupling effect becomes significant, and the results show a drastic difference between the center and outer fingers (e.g., the center finger becomes much hotter than the outer fingers; see Figures 4.8(c, d)). The center finger will eventually conduct the most current, and the other two fingers become nearly inactive. This leads to a sharp decrease in the collector current, which is a phenomenon called thermal runaway.

Figure 4.9(a) shows the Gummel plot calculated from the model and obtained from measurement for a 6-finger HBT at $V_{CB} = 0$V, 3V, and 6V. The HBT has a typical intrinsic make-up of $N_E = 5 \times 10^{17}$ cm$^{-3}$, $N_B = 8 \times 10^{18}$ cm$^{-3}$, $N_C = 5 \times 10^{16}$ cm$^{-3}$, emitter layer thickness of 1000 Å, base layer thickness of 1000 Å, collector thickness of 7000 Å, a finger area of $2.5 \times 10$ µm$^2$, and a ballast emitter contact resistance of $6 \times 10^{-6}$ Ω-cm$^2$. The extrinsic make-up of the HBT, which is needed to calculate the thermal resistance $R_{th}$, is $Z = 25$ µm, $A_C = 10 \times 25$ µm$^2$, and $X_s = 100$ µm. This gives a thermal resistance of about 400 K/W at room temperature. As shown in the figure, both the collector and base currents ($I_C$ and $I_B$) increase as $V_{CB}$ is increased due to the fact that the heat power generated, and therefore the temperature, in the HBT increase as $V_{CE}$ ($V_{CE} = V_{BE} + V_{CB}$) increases. But since $I_B$ rises more quickly than $I_C$, the current gain will decrease as $V_{CE}$ is increased when the current level is high. This trend is clearly shown in Figure 4.9(b).

The model developed can also be used to calculate $I_C$ versus $V_{CE}$ characteristics for constant $I_B$. The results, together with experimental data, are given in Figure 4.10. A negative slope on the $I_C$-$V_{CE}$ characteristics is observed when the base current is large where the thermal effect becomes prominent. This can be attributed to the uneven increase of $I_C$ and $I_B$ as $V_{CB}$, or $V_{CE}$, is increased (see Figure 4.9(a)). The increased $I_B$ due to the thermal effect reduces the base-emitter voltage $V'_{BE}$ required to maintain that constant base current, which subsequently decreases the collector current. The values of $V'_{BE}$ corresponding to the $I$-$V$ curves in Figure 4.10 are shown in Figure 4.11, which indicate that the $V'_{BE}$ needed to maintain the constant $I_B$ is almost constant when the base current is low and decreased rapidly versus $V_{CE}$ when the base current is high.

Note that the experimental data shown in Figure 4.10 increases slightly at relatively low $I_B$ and large $V_{CE}$. This is caused by avalanche multiplication in the reverse-biased base-collector junction—an effect not accounted for in the present model. It is interesting to see that the elevated temperature at higher $I_B$ seems to offset the avalanche effect and subsequently increase the breakdown voltage.

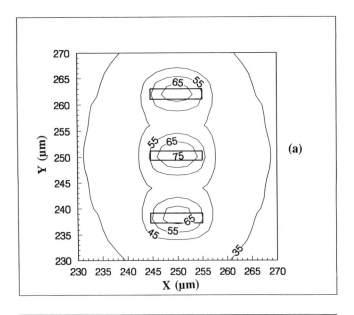
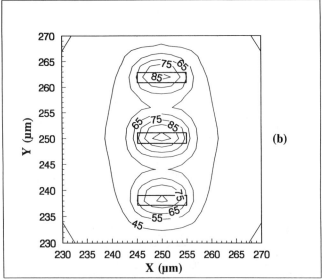

**Figure 4.8** Contour plots of the temperature distribution at the surface of the 3-finger HBT biased at $I_B = 2$ mA and at (a) $V_{CE} = 2$V, (b) $V_{CE} = 3$V, (c) $V_{CE} = 5$V, and (d) $V_{CE} = 6$V.

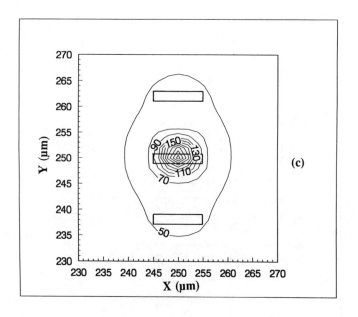

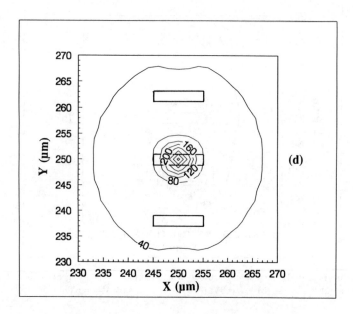

**Figure 4.8** (continued)

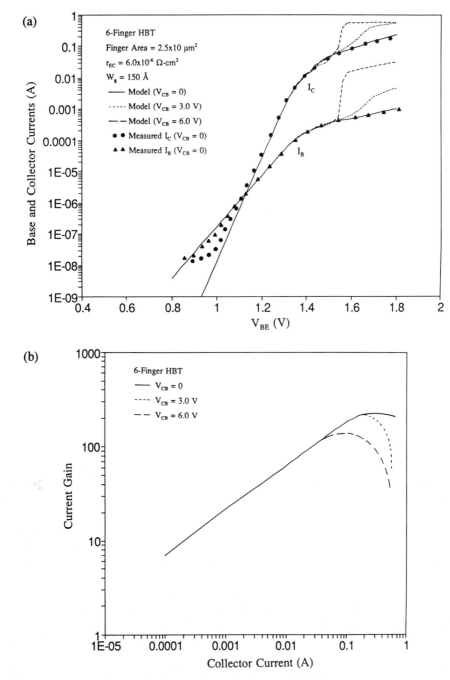

**Figure 4.9** (a) Base and collector currents calculated from the present model for a 6-finger HBT at $V_{CB} = 0$V, 3V, and 6V. Also included are the experimental data measured at $V_{CB} = 0$. (b) dc current gain corresponding to the results shown in Figure 4.9(a).

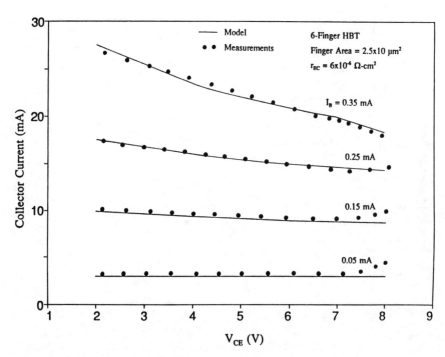

**Figure 4.10** Collector current versus collector-emitter voltage characteristics as a function of constant $I_B$ for a 6-finger HBT with ballast emitter resistance.

Even with a ballast resistance, thermal runaway can still prevail but is less obvious and occurs only if $V_{CE}$ is sufficiently large. This is evidenced by the "minor thermal runaway" at $I_B = 0.35$ mA and $V_{CE} \geq 7$V shown in Figure 4.10. The thermal runaway results from the even more asymmetrical increase in the base and collector currents at a higher base current as seen in Figure 4.9(a) (at $V_{BE} = 1.45$V and $V_{CB} = 6$V). If $I_B$ is fixed, then the voltage $V'_{BE}$ required to maintain that $I_B$ will also decrease sharply when $V_{CE}$ is increased beyond a critical value (see Figure 4.11), which then sharply decreases the collector current. As will be shown later, the thermal runaway is much more apparent if the HBT does not have a ballast emitter resistance.

We next examine the effect of the emitter contact resistance $r_{EC}$ on the HBT performance. Here we consider a 3-finger HBT. Figure 4.12(a) shows the Gummel plot for three different $r_{EC}$ and $V_{CB} = 2$V. The same plot for a larger $V_{CB}$ (= 5V) is given in Figure 4.12(b). Obviously, the increased $r_{EC}$ suppresses both the collector and base currents at high $V_{BE}$. Of equal importance to note is the behavior of the collector current at large $r_{EC}$ and when $V_{CE}$ is high. Let us use the results in Figures 4.12(a, b) to illustrate this point. First consider the nonballast $r_{EC}$ case ($r_{EC} = 10^{-6}$ Ω/cm²) and fix $I_B = 1$ mA. When

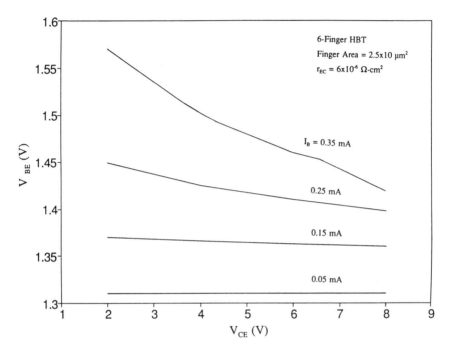

**Figure 4.11** The base-emitter voltage $V'_{BE}$ that is required to maintain the constant $I_B$ used in Figure 4.9.

$V_{CB} = 2V$ (or $V_{CE} \approx 3V$), $V'_{BE} \approx 1.62V$ and $I_C \approx 0.2A$. As $V_{CB}$ is increased to 5V ($V_{CE} \approx 6V$), $V'_{BE} \approx 1.44V$, and the corresponding $I_C$ is about 0.1A, which was decreased to about 50 percent of its previous value (thermal runaway occurs).

Now let us consider the HBT with a ballast resistance ($r_{EC} = 2.5 \times 10^{-6}$ $\Omega/cm^2$) and fix $I_B$ at 0.3 mA (the maximum value shown in Figure 4.12(a)). At $V_{CE} = 3V$, the corresponding $I_C$ is 0.04A; at $V_{CE} = 6V$, $I_C$ is about 0.035A. Thus, the collector current in this case is decreased only slightly as $V_{CE}$ is increased from 2V to 6V.

The emitter-finger temperatures corresponding to the results in Figure 4.12(b) are shown in Figure 4.13. Note that the temperature at the center finger is higher than that at the outer fingers, which arises from the fact that the center finger is subjected to two thermal coupling whereas the outer fingers are subjected to only one thermal coupling.

The thermal runaway phenomenon in a multifinger HBT without the ballast emitter resistance discussed above can be clearly illustrated by the I-V characteristics shown in Figure 4.14. The device considered has 4 emitter fingers, a finger area of $2.5 \times 20$ mm$^2$, and a low emitter contact resistance ($r_{EC} \approx 10^{-6}$ $\Omega/cm^2$). The onset of thermal runaway is observed at $V_{CE} \approx 5V$ when $I_B$ is increased beyond 2.5 mA.

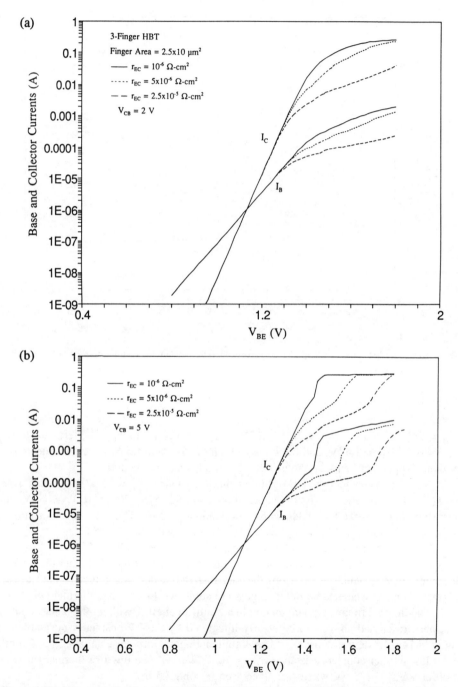

**Figure 4.12** Base and collector currents calculated from the present model for three different emitter contact resistances at (a) $V_{CB} = 2V$ and (b) $V_{CB} = 5V$.

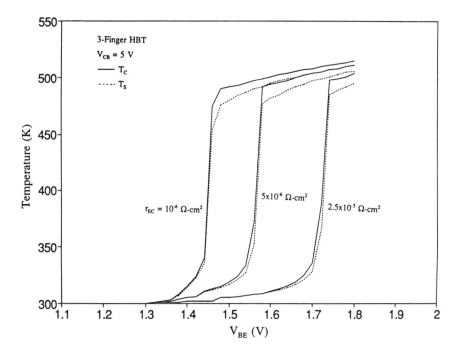

**Figure 4.13** Temperatures at the center and outer fingers for the device considered in Figure 4.11 and $V_{CB} = 5V$.

## 4.3 THERMAL-AVALANCHE INTERACTING BEHAVIOR

Avalanche multiplication is an important mechanism for bipolar transistors because it imposes an upper limit of the collector-emitter voltage $V_{CE}$ for a given base current [19]. In an AlGaAs/GaAs HBT, avalanche behavior is more complex than that in a silicon bipolar transistor due to the significance of the self-heating effect at high $V_{CE}$ caused by the poor thermal conductivity of GaAs [15]. Furthermore, for a multiple emitter finger HBT used frequently in microwave applications, the thermal-coupling effect among the neighboring fingers also plays an important role in the HBT thermal-electrical characteristics (see Section 4.2). The interaction of the thermal and avalanche mechanisms is therefore a vital issue in understanding the HBT behavior operating at the high-power region.

This section presents an analytical model to describe the HBT thermal-avalanche interacting behavior. The commonly used multiple emitter finger HBT structure is considered, and an empirical model including self-heating and thermal-coupling effects as well as the impact ionization mechanism is developed. Experimental data measured from a 6-finger, high-power HBT will be included to support the model.

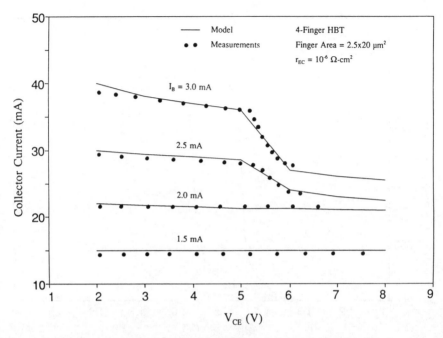

**Figure 4.14** Collector current versus collector-emitter voltage characteristics as a function of constant $I_B$ for a 4-finger HBT without ballast emitter resistance.

When the lattice temperature in the HBT is increased, which can result from a high current and/or high voltage, it increases the base current and, to a lesser extent, the collector current [18]. As a result, when $I_B$ is fixed and $V_{CE}$ is increased, which increases the heat power generated and thus the lattice temperature in the HBT, the base-emitter voltage $V'_{BE}$ required to maintain such a particular $I_B$ is reduced. This reduced $V'_{BE}$ subsequently reduces the collector current $I_C$. Therefore, the HBT $I_C$-$V_{CE}$ characteristics exhibit a negative slope where the thermal effect becomes important. On the other hand, $I_C$ tends to increase sharply at large $V_{CE}$ due to the effect of impact ionization in the reverse-biased base-collector junction [20].

Consider a particular emitter finger (called subject finger) in the multifinger HBT. We propose the following empirical expression that fits the experimental results:

$$I_C^{k+1} = I_C^k M^{k+1} (T^{k+1}/T_0)^{-A_1} \qquad k = 0, 1, 2, \ldots, N \qquad (4.18)$$

where superscript $k$ is the number of $V_{CE}$ increments (e.g., $k = 0$ and $k = 1$ represent the first and second $V_{CE}$ points used in calculations, respectively), $M$ is the avalanche multiplication factor, $T$ is the HBT lattice temperature, $T_0$ (= 300K) is the ambient temperature, and $A_1$ is an empirical parameter accounting for the self-heating effect ($A_1 = 0$ if the self-heating effect is neglected). The term $M$ accounts for the avalanche multiplication effect [20]; the physical mechanisms responsible for the negative conductance, modeled by $(T/T_0)^{-A_1}$ in (4.18), were discussed briefly in the beginning of this chapter and in more detail in [18]. Note that the initial $I_C$ ($I_C^0$) should be taken at a very small $V_{CE}$ (e.g., $V_{CE}^0 = 1V$). Thus, $T^0 \approx T_0$ and $M^0 = 1$, and the value of $I_C^0$ for a given $I_B$ can be determined from a conventional HBT model [21] without concern about the thermal and avalanche effects.

For a known $I_C^k$, $T^{k+1}$ on the subject finger can be determined as [13, 22]

$$T^{k+1} = T_0 + P_S^{k+1} R_S^{k+1} + P_C^{k+1} + R_C^{k+1} \quad (4.19)$$

where $P_S$ and $P_C$ are the self-heating powers generated on the subject finger and on the neighboring finger(s), respectively; and $R_S$ and $R_C$ are the thermal resistances of the subject finger due to self-heating and due to thermal-coupling between the subject and neighboring fingers, respectively. $P$ and $R_C$ are given by

$$P^{k+1} = I_C^k V_{CE}^{k+1} \quad \text{and} \quad R_C = A_2 R_S \quad (4.20)$$

where $A_2$ is another empirical parameter that accounts for the thermal-coupling effect. The value of $A_2$ depends strongly on the emitter-finger geometry, finger spacing, and metalization process ($A_2 = 0$ for a single-finger HBT or a multifinger HBT with very large finger spacing). Assuming that the heat in the HBT is dissipated primarily through the substrate yields

$$R_S = \int_0^{X_s} dx/\{k_S[A_C + 2Z \tan(\theta)x]\} \quad (4.21)$$

Here $X_s$ is the substrate thickness, $A_C$ is the collector area, $Z$ is the HBT width, $\theta = 45°$ is the lateral heat diffusion angle, and $K_s = 0.47(T/T_0)^{-1.22}$ is the GaAs thermal conductivity [11].

The avalanche multiplication factor $M$ is given by [20]

$$M = 1 + \int_0^X \alpha \exp\{-[b/\xi(x)]^\beta\} \, dx \quad (4.22)$$

where $0 < x < X$ is the base-collector space-charge region, a is the impact ionization coefficient, $b$ is the critical field, $\beta$ is the exponential factor, and $\xi(x)$ is the position-dependent electric field in the space-charge region [20]. The temperature-dependent impact ionization parameters for GaAs are [23]

$$\alpha = \alpha_{300}[1 + c(T - T_0)]\xi(x) \quad (4.23)$$

$$b = b_{300}[1 + d(T - T_0)] \quad (4.24)$$

and $\beta = 2.0$. Here $\alpha_{300} = 0.294$ V$^{-1}$, $b_{300} = 5.86 \times 10^5$ Vcm$^{-1}$, $c = 8.5 \times 10^{-4}$ K$^{-1}$, and $d = 7.17 \times ^{-4}$ K$^{-1}$. Note that $\xi(x)$ and $X$, and thus $M$, are influenced strongly by the collector current passing through the space-charge region. An extreme case is the occurrence of current-induced base pushout at a sufficiently high current level [24]. It can be seen from (4.23) and (4.24) that both $\alpha$ and $b$ will increase if $T$ is increased. However, since $b$ affects $M$ more strongly than $\alpha$, $M$ will decrease as $T$ increases. The underlying physics for this occurrence is that the electrons cannot gain enough kinetic energy for impact ionization due to enhanced lattice scattering at elevated temperatures [23].

After $T^{k+1}$ and $M^{k+1}$ are found, $I_C^{k+1}$ for the subject finger can be calculated from (4.18). Since the thermal and avalanche effects interact with each other, a simple iteration procedure is needed to obtain a self-consistent solution of the temperature and collector current at each finger as a function of $V_{CE}$.

Figure 4.15 compares the $I$-$V$ characteristics obtained from the thermal-avalanche

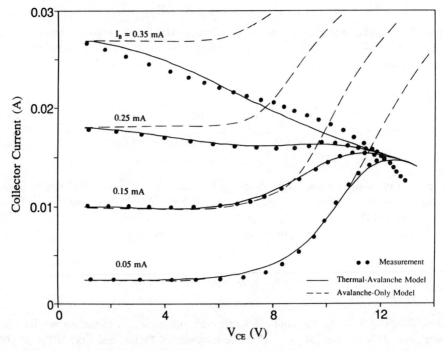

**Figure 4.15** $I$-$V$ characteristics calculated from the thermal-avalanche model, calculated from the avalanche-only model, and obtained from measurements for the 6-finger HBT.

model, avalanche-only model, and measurements for a six-finger HBT. The device was fabricated on an MOCVD-grown structure with a $2.5 \times 10$ μm² finger area; 10-μm finger spacing; and an epi structure that includes $5 \times 10^{17}$ cm$^{-3}$ emitter, $5 \times 10^{19}$ cm$^{-3}$ base, and $3 \times 10^{16}$ cm$^{-3}$ collector doping concentrations and 1700-Å emitter, 1000-Å base, and 7000-Å collector layer thicknesses. The calculated thermal resistance associated with the substrate is about 480 K/W at room temperature, compared to 400 K/W from measurement. The empirical parameters $A_1 = 0.03$ and $A_2 = 0.5$ are used in calculations, which are determined by fitting the model results with experimental data. Clearly, thermal effect suppresses avalanche multiplication at high values of $V_{CE}$ due to the fact that a high $V_{CE}$ gives rise to a high lattice temperature, which then reduces the multiplication factor $M$, as can be seen in Figure 4.16 by comparing $M$ calculated from the thermal-avalanche and avalanche-only models at $I_B = 0.05$ mA. The trend is different, however, at $I_B = 0.35$ mA (Figure 4.16).

Let us first consider the case of thermal-avalanche model for $I_B = 0.05$ mA. For $V_{CE} < 10$V, $M$ increases because an increased $V_{CE}$ gives rise to a larger electric field in the space-charge region. As $V_{CE}$ increase beyond 10V, avalanche multiplication occurs and the current and lattice temperature are increased significantly. Such increases thus induce base pushout, which subsequently reduces the base-collector depletion layer thickness

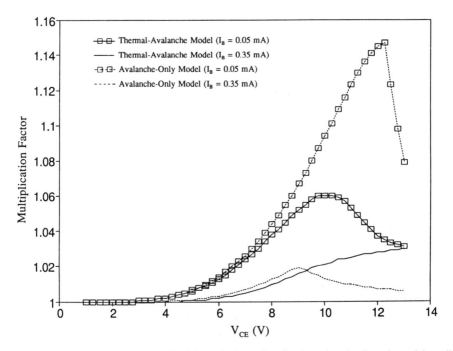

**Figure 4.16** Multiplication factors calculated from the thermal-avalanche and avalanche-only models at different base currents.

(note that the entire epi-collector becomes the depletion region before the onset of base pushout, and the quasi-neutral region pushes into the epi-collector when base pushout occurs) and therefore reduces the multiplication factor if $V_{CE}$ is further increased.

The multiplication factor calculated from the thermal-avalanche model for $I_B = 0.35$ mA shows a different behavior and increases monotonically versus $V_{CE}$. We suggest this result based on the fact that the base pushout has occurred at very low $V_{CE}$ due to the high $I_C$ and the high lattice temperature. However, unlike the previous cases, the electric field increase is now more important than the depletion layer thickness reduction due to the suppressed $I_C$ at the elevated temperature, which gives rise to the increasing $M$ characteristics.

The temperatures at the center two fingers (hottest fingers among the six fingers due to thermal coupling) are plotted in Figure 4.17. Temperatures at the outer and outermost fingers are about 80 percent and 60 percent of those at the center fingers. Note that the measured $I_C$-$V_{CE}$ characteristics (Figure 4.15) exhibit a sharper decrease than the model predictions at $V_{CE} > 11$V. This can be attributed to thermal runaway at very high temperatures—an effect accounted for in the present model only in a first-order manner and the exact solution of which requires solving the three-dimensional heat transfer equation [13]. Also, it should be pointed out that second breakdown, which can occur at even higher $V_{CE}$, is not treated in the present model and may form a subject for future research.

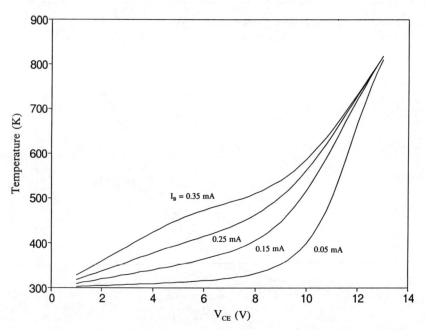

**Figure 4.17** Temperatures at the center two fingers of the 6-finger HBT calculated versus $V_{CE}$ and as a function of $I_B$.

## 4.4 HBTs OPERATING BETWEEN 300K AND 500K

The AlGaAs/GaAs HBT has emerged as a prominent and practical device for high-speed and high current-handling applications. Most studies on HBTs reported to date [25], including those given in the previous sections, focus on the HBT performance at room temperature. Recently, there has been a rapid growth of HBT circuits designed to operate in the 300-K to 500-K temperature range [26]. As a result, analysis of high temperature HBT performance becomes increasingly important. Some efforts on understanding the high-temperature HBT behavior have been reported [27, 28], but a detailed theoretical study that can provide the physical insight of such a subject is not yet available.

In this section, the dc performance of a single-emitter finger HBT operating at high ambient temperatures (between 300 K and 500 K) is investigated. Both the large- and small-area HBTs are considered. The major difference between the two HBTs is that the self-heating effect is less significant in the former device because of its large area, which allows the heat generated in the HBT to be dissipated quickly.

The model and procedure developed in Section 4.1 are still applicable here for calculating the HBT base and collector currents, including the self-heating effect, at higher ambient temperatures, provided the ambient temperature $T_0$ is now larger than 300K. Since the lattice temperature $T$ and the currents of the HBT are interacting with each other, a simple iteration procedure is required to solve $T$, $J_C$, and $J_B$ [7]. First, the currents are calculated using a first-order $T$, which is normally chosen as the ambient temperature $T_0$. Such currents are then used to find the heat power generated in the HBT and the second-order $T$. The iteration continues until the solutions are converged.

### 4.4.1 Small-Area HBT

We first consider a small-area ($2 \times 10$ μm$^2$) HBT structure depicted in Table 4.2. Figure 4.18(a) shows the Gummel plot of the HBT operated at ambient temperatures $T_0 = 300$K and 500K. The results are calculated from the present model and simulated from a two-dimensional device simulator MEDICI [29] excluding the self-heating effect (i.e., $T = T_0$). In addition to the ability of simulating heterostructure semiconductor devices, MEDICI allows the effects of lattice heating to be included in simulation by solving the

**Table 4.2**
AlGaAs/GaAs HBT Structure Considered for 300-K to 500-K Operation

|  | Material | Thickness (Å) | Composition | Doping (cm$^{-3}$) |
|---|---|---|---|---|
| Emitter | n − Al$_x$Ga$_{1-x}$As | 1000 | $x = 0.3$ | $5.0 \times 10^{17}$ |
| Base | p − GaAs | 1000 |  | $1.0 \times 10^{19}$ |
| Collector | n − GaAs | 6000 |  | $5.0 \times 10^{16}$ |

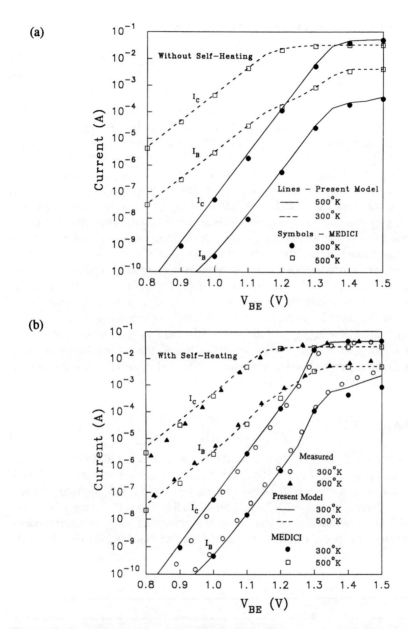

**Figure 4.18** Gummel plots of the small-area HBT calculated from the present model (a) without and (b) with the self-heating effect. Also included are the results simulated from the MEDICI two-dimensional device simulator and obtained from measurements (symbols).

heat transfer equation. For example, to compute the spatially dependent lattice temperature $T$, the heat flow equation is used [29].

$$\delta c \partial T/\partial t = H + \nabla[\lambda(T)\nabla T] \quad (4.25)$$

where $\delta$ is the mass density, $c$ is the specific heat, $H$ is the heat generation, and $\lambda$ is the thermal conductivity of the material. Detailed discussions on MEDICI simulation of AlGaAs/GaAs HBTs will be given in Chapter 7. Results calculated from the model and simulated from MEDICI with the self-heating effect are presented in Figure 4.18(b). Also included in the figure are measurement data.

Comparing the results of Figures 4.18(a, b), we find that $I_C$ and $I_B$ at small $V_{BE}$ calculated without the self-heating effect are identical to those calculated with the self-heating effect because lattice heating is directly proportional to the current level and therefore is insignificant when the current is small. For large $V_{BE}$, however, both $I_C$ and $I_B$ are increased by the self-heating effect, but with a larger increase in $I_B$ than in $I_C$. Such an uneven increase thus leads to a current gain degradation at high currents, as illustrated by the results shown in Figure 4.19. Note that the current gains calculated with and without the self-heating effect start to deviate at a smaller current as the ambient

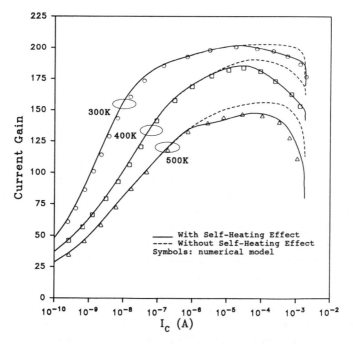

**Figure 4.19** Current gain versus $I_C$ calculated and simulated with self-heating effect at three different ambient temperatures.

temperature is increased from 300K to 500K, indicating that the self-heating effect is more prominent at higher ambient temperatures.

Figures 4.20(a, b) illustrate the three-dimensional lattice temperature contours in the HBT operating at ambient temperatures of 300K and 500K, respectively. In the figure, the HBT layout is in the $x$-$y$ space ($y = 0$ is the surface the HBT), and the emitter finger is located at 249 μm $< x <$ 251 μm. These results are simulated at $V_{BE} = 1.5$V and $V_{CE} = 3$V. Clearly, the peak temperature occurs in the intrinsic HBT (including emitter, base, and collector layers, with a total thickness of $0 < y < 1.15$ μm), and the lattice temperature decreases rapidly toward the side-wall and semi-insulating substrate ($y > 1.15$ μm). Also note that the self-heating effect is more significant as the ambient temperature is increased (e.g., from 300K to 360K in Figure 4.20(a) and from 500K to 650K in Figure 4.20(b)).

### 4.4.2 Large-Area HBT

Here, we consider an HBT that has a larger emitter area (100 × 100 μm$^2$) than the HBT considered in the previous section. Such a device permits a rapid heat dissipation, thus allowing us to neglect the self-heating effect and to study solely the effects of higher ambient temperature on the HBT current characteristics. Figure 4.21 compares the collector current $I_C$ calculated from the present model (solid lines) and obtained from MEDICI simulation (symbols). When the ambient temperature is increased, $I_C$ increases but saturates at a smaller $V_{BE}$ due to the reduction of the barrier potentials at the emitter-base heterojunctions as the temperature is increased, which enhances the free carrier injection from the emitter to base.

Figure 4.22 shows the base current versus $V_{BE}$ characteristics. Like the collector current, $I_B$ increases with increasing temperature, but the degree of $I_B$ increase is larger than that found in Figure 4.21. Such an increase is again a result of barrier potentials being reduced at higher ambient temperatures.

It is worth emphasizing that while both $I_C$ and $I_B$ are subjected to the same barrier potential decrease as the ambient temperature increases, their increases are governed by different physical mechanisms. The free-carrier transport from the emitter to base is controlled by the thermionic and tunneling mechanisms due to the presence of the spike at the heterointerface. On the other hand, the hole current injected from the base to emitter is predominantly carried out by drift and diffusion tendencies. Numerous studies have shown that the drift and diffusion mechanism yields a higher injection efficiency than its thermionic and tunneling counterpart [30, 31].

The uneven increases in $I_C$ and $I_B$, as just discussed, lead to a reduction of the current gain β at higher temperatures, as evidenced by the results given in Figure 4.23. Experimental data reported in [27] are also included in the figure (symbols). Note that the HBT has nearly flat current gains at high current levels for all three temperatures, indicating that the self-heating effect is not significant in the large-area HBT.

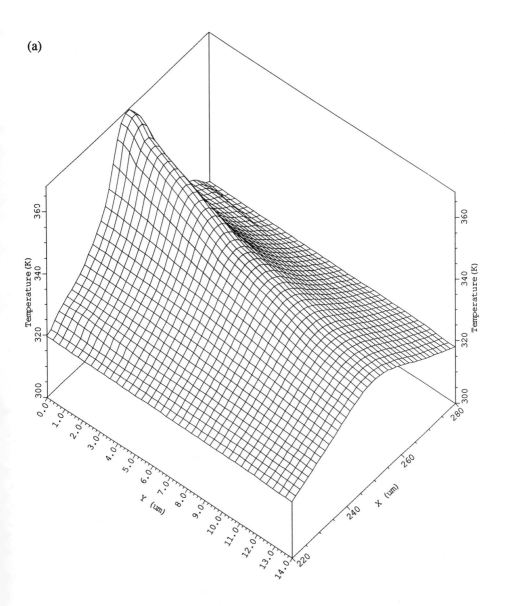

**Figure 4.20** Three-dimensional lattice temperature contours in the small-area HBT operating at ambient temperatures of (a) 300K and (b) 500K.

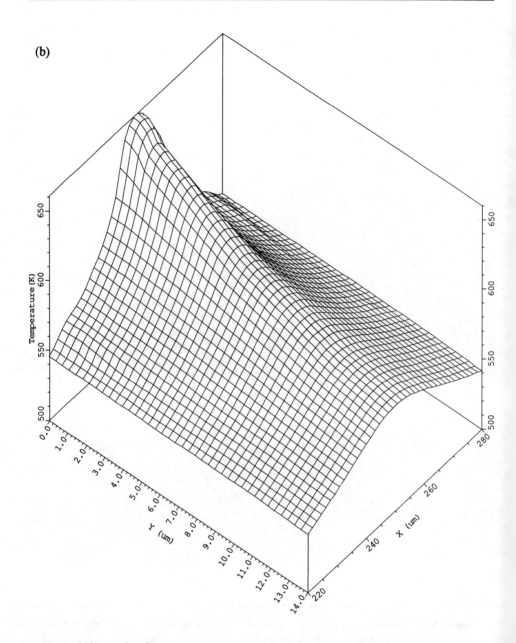

**Figure 4.20** (continued)

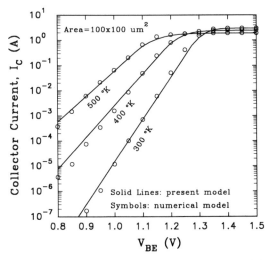

**Figure 4.21** Collector current $I_C$ versus $V_{BE}$ of the large-area HBT calculated and simulated at three different ambient temperatures.

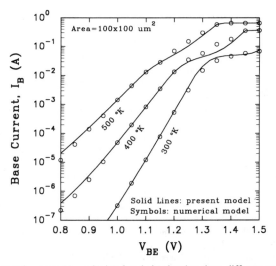

**Figure 4.22** Base current $I_B$ versus $V_{BE}$ calculated and simulated at three different ambient temperatures.

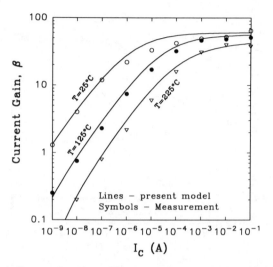

**Figure 4.23** Current gain β versus $I_C$ at three different temperatures calculated from the present model and obtained from measurements (*Source*: [27]. © 1992 IEEE.).

# References

[1] Kim, M. E., B. Bayraktaroglu, and A. Gupta, "HBT Devices and Applications," in *HEMTs & HBTs: Devices, Fabrication, and Circuits*, edited by F. Ali and A. Gupta, Norwood: Artech House Inc., 1991.

[2] Wang, N. L., N. H. Sheng, M. F. Chang, W. J. Ho, G. J. Sullivan, E. A. Sovero, J. A. Higgins, and P. M. Asbeck, "Ultrahigh Power Efficiency Operation of Common-Emitter and Common-Base HBTs at 10 GHz," *IEEE Trans. Microwave Theory and Tech.*, Vol. 38, 1990, pp. 1381–1389.

[3] Khatibzadeh, M. A., B. Bayraktaroglu, and T. Kim, "12 W Monolithic X-Band HBT Power Amplifier," *IEEE MTT-S Int. Microwave Symp. Digest*, 1992, pp. 47–50.

[4] Liou, L. L., B. Bayraktaroglu, and C. I. Huang, "Thermal Stability Analysis of Multiple Finger Microwave AlGaAs/GaAs Heterojunction Bipolar Transistor," *IEEE Int. Microwave Symp. Tech. Digest*, 1993.

[5] Gao, G. B., M. S. Unlu, H. Morkoc, and D. L. Blackburn, "Emitter Ballasting Resistor Design for Current Handling Capability of AlGaAs/GaAs Power Heterojunction Bipolar Transistors," *IEEE Trans. Electron Devices*, Vol. 38, 1991, pp. 185–196.

[6] Liou, J. J., *Advanced Semiconductor Device Physics and Modeling*, Norwood: Artech House, 1994, Chapter 1.

[7] Shur, M., *Physics of Semiconductor Devices*, Englewood Cliffs, NJ: Prentice-Hall, 1990.

[8] Chen, S.-C., Y.-K. Su, and C.-Z. Lee, "A Study of Current Transport on p-N Heterojunctions," *Solid-St. Electron.*, Vol. 35, 1992, p. 1311.

[9] Sunderland, D. A., and P. L. Dapkus, "Optimizing N-p-n and P-n-p Heterojunction Bipolar Transistors for Speed," *IEEE Trans. Electron Devices*, Vol. ED-34, 1987, pp. 367–377.

[10] Maycock, D. P., "Thermal Conductivity of Silicon, Germanium, III-V Compound and III-V Alloys," *Solid-St. Electron.*, Vol. 10, 1967, p. 161.

[11] Joyce, W. B., "Thermal Resistance of Heat Sink with Temperature-Dependent Conductivity," *Solid-St. Electron.*, Vol. 18, 1975, p. 321.

[12] Meeks, D., "Fundamentals of Heat Transfer in a Multilayer System," *Microwave J.*, January 1992, pp. 165–172.

[13] Liou, L. L., C. I. Huang, and J. Ebel, "Numerical Studies of Thermal Effects on Heterojunction Bipolar Transistor Current-Voltage Characteristics Using One-Dimensional Simulation," *Solid-St. Electron.*, Vol. 35, 1992, pp. 579–585.
[14] Zheng, L. R., S. A. Wilson, D. J. Lawrence, S. I. Rudolph, S. Chen, and G. Braunstein, "Shallow Ohmic Contacts to n-Type GaAs and $Al_xGa_{1-x}As$," *Appl. Phys. Lett.*, Vol. 60, 1992, pp. 877–879.
[15] Gao, G. B., M. Z. Wang, X. Gui, and H. Morkoc, "Thermal Design Studies of High-Power Heterojunction Bipolar Transistors," *IEEE Trans. Electron Devices*, Vol. ED-36, 1989, p. 854.
[16] Whitefield, D. S., C. J. Wei, and J. C. M. Hwang, "Temperature-Dependent Large-Signal Model of Heterojunction Bipolar Transistors," *IEEE GaAs IC Symp. Tech. Digest*, 1992, pp. 221–224.
[17] Marty, A., T. Camps, J. Tasselli, D. L. Pulfrey, and J. P. Bailbe, "A Self-Consistent dc-ac Two-Dimensional Electrothermal Model for GaAlAs/GaAs Microwave Power HBT's," *IEEE Trans. Electron Devices*, Vol. 40, 1993, p. 1202.
[18] Liou, J. J., L. L. Liou, C. I. Huang, and Bayraktaroglu, "A Physics-Base Heterojunction Bipolar Transistor Including Thermal and High-Current Effects," *IEEE Trans. Electron Devices*, Vol. 40, 1993, p. 1570.
[19] Lu, P.-F., and T.-C. Chen, "Collector-Base Junction Avalanche Effects in Advanced Double-Poly Self-Aligned Bipolar Transistors," *IEEE Trans. Electron Devices*, Vol. 36, 1989, p. 1182.
[20] Liou, J. J., *Advanced Semiconductor Device Physics and Modeling*, Norwood: Artech House, Inc., 1994, Chapter 3.
[21] Grinberg, A. A., M. S. Shur, R. J. Fischer, and H. Morkoc, "An Investigation of the Effect of Graded Layers and Tunneling on the Performance of AlGaAs/GaAs Heterojunction Bipolar Transistors," *IEEE Trans. Electron Devices*, Vol. ED-31, 1984, p. 1758.
[22] Liu, W., and B. Bayraktaroglu, "Theoretical Calculations of Temperature and Current Profiles in Multi-finger Heterojunction Bipolar Transistors," *Solid-St. Electron.*, Vol. 36, 1993, p. 125.
[23] Okuto, Y., and C. R. Crowell, "Threshold Energy Effect on Avalanche Break-Down Voltage in Semiconductor Junctions," *Solid-St. Electron.*, Vol. 18, 1975, pp. 161–168.
[24] Bowler, D. L., and F. A. Lindholm, "High Current Regimes in Transistor Collector Regions," *IEEE Trans. Electron Devices*, Vol. ED-20, 1973, p. 257.
[25] Tiwari, S., and D. J. Frank, "Analysis of the Operation of GaAlAs/GaAs HBT's," *IEEE Trans. Electron Devices*, Vol. 36, 1989, p. 2105.
[26] Dikmen, C. T., and N. S. Dogan, "Modeling and Characterization of AlGaAs/GaAs Heterojunction Bipolar Transistors for High-Temperature Applications," *Trans. 2nd Int. High Temperature Electronics Conf.*, June 1994, pp. P51–P56.
[27] Fricke, K., H. L. Hartnagel, W. Y. Lee, and J. Wurfl, "AlGaAs/GaAs HBT for High-Temperature Applications," *IEEE Trans. Electron Devices*, Vol. 39, 1992, pp. 1977–1981.
[28] Liu, W., S. K. Fan, T Henderson, and D. Davito, "Temperature Dependences of Current Gains in GalnP/GaAs and AlGaAs/GaAs Heterojunction Bipolar Transistors," *IEEE Trans. Electron Devices*, Vol. 40, July 1993, pp. 1351–1353.
[29] *MEDICI Manual*, Technology Modeling Associates, Inc., Palo Alto, CA, 1993.
[30] Das, A., and M. S. Lundstrom, "Numerical Study of Emitter-Base Junction Design for AlGaAs/GaAs Heterojunction Bipolar Transistors," *IEEE Trans. Electron Devices*, Vol. 35, 198, p. 863.
[31] Chen, S.-C., Y.-K. Su, and C.-Z. Lee, "A Study of Current Transport on P-N Heterojunctions," *Solid-St. Electron.*, Vol. 35, 1992, p. 1311.

# Chapter 5
# Base and Collector Leakage Currents of HBTs

So far we have developed various models to describe the AlGaAs/GaAs HBT performance operating at normal bias conditions (e.g., relatively large applied base-emitter voltage). However, for HBTs subjected to small bias voltages, base and collector leakage currents are often the dominant components of the base and collector currents, respectively. Such currents have been widely observed in experimental measurements, and a good understanding of their origin and physical mechanisms will be highly useful to HBT design and characterization. Furthermore, the leakage currents often serve as useful indicators to the quality of the HBT peripheral surface properties and therefore have important implications to the long-term stability of HBT. Despite the extensive studies of the other current components, such as the space-charge region recombination current [1], surface recombination current [2,3], and base bulk recombination current [4], the properties of leakage currents have largely been overlooked in the literature.

In this chapter, the physical mechanisms underlying the base and collector leakage currents of AlGaAs/GaAs HBTs are discussed, and an analytical model is developed to characterize the leakage current behavior. The relevance of leakage currents to the HBT reliability issue is also addressed.

The structure of a typical mesa-etched, single-emitter finger HBT is shown in Figure 5.1. The base and collector leakage currents can take place at the emitter-base perimeter (circle 1 in Figure 5.1), base-collector perimeter (circle 2), collector-subcollector perimeter (circle 3), and subcollector-substrate interface (circle 4). The magnitude of such currents depends strongly on the etching process, the quality of the emitter-base and base-collector peripheries that are covered by the dielectric (e.g., polyimide and nitride) layer, and the quality of the $n^+$-GaAs/semi-insulating-GaAs interface. For discussions, we assume the HBT surface is protected with a polyimide layer. Inferior-quality emitter and base peripheries can thus increase the possibility for the free carriers either to surmount or to tunnel through the GaAs-polyimide and AlGaAs-polyimide interface barriers and subsequently increase the leakage currents [5]. As will be shown later, the peripheral

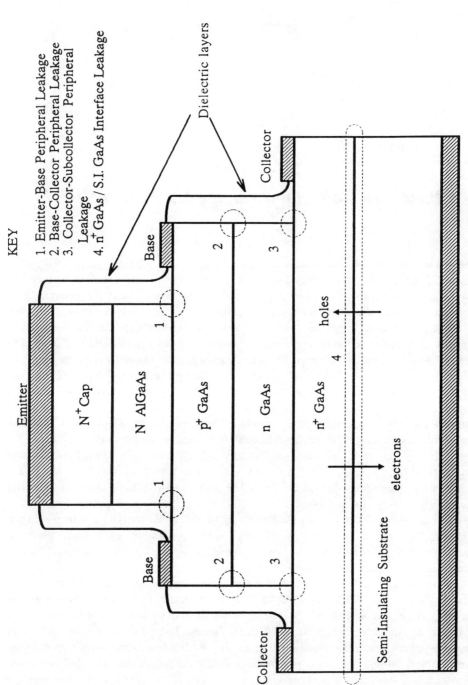

**Figure 5.1** Schematic illustration of the cross section of a typical HBT structure, with leakage peripheries circled.

quality is insensitive to the type of dielectric layer but rather varies strongly from process to process.

## 5.1 LEAKAGE CURRENTS AT THE EMITTER-BASE PERIPHERY

The base and collector leakage currents that originate at the emitter perimeter ($I_{BL,E}$ and $I_{CL,E}$) are affected by the applied base-emitter voltage $V_{BE}$. As $V_{BE}$ is increased, the potential barriers at the GaAs-polyimide and AlGaAs-polyimide interfaces are lowered, thus increasing the numbers of electrons in the n-type AlGaAs and holes in the p-type GaAs to surmount the barriers and to reach the base and emitter regions, respectively. Once there, *these excess minority carriers are no different from those injected across the emitter-base heterojunction and thereby contribute additional base and collector currents.*

The leakage currents can be modeled using the physical degradation mechanism described in the Arrhenius relationship [6], which expresses the physical-chemical reaction rate as

$$k_1 = k_{10}\exp(-E_a/kT) \tag{5.1}$$

where $k_1$ is the reaction rate, $k_{10}$ is a constant peculiar to the reaction type, $E_a$ is the activation energy of the degradation process, $k$ is the Boltzmann constant, and $T$ is the absolute temperature. This relation suggests that the degradation rate decreases sharply as $E_a$ is increased. The total activation rate $k_t$ is obtained by integrating the right-hand side of (5.1) from $E_a = 0$ to the highest activation energy $E'_a$ as

$$k_t = k'[1 - \exp(-E'_a/kT)] \tag{5.2}$$

where $k'$ is a constant related to $k_{10}$.

Based on the above concept, $I_{BL,E}$ and $I_{CL,E}$ can be expressed analogous to (5.2) as

$$I_{BL,E} = NP_E J_{BL,E} = NP_E J'_{BL,E}[1 - \exp(-V_{BE}F_L/V_T)] \tag{5.3}$$

$$I_{CL,E} = NP_E J_{CL,E} = NP_E J'_{CL,E}[1 - \exp(-V_{BE}F_L/V_T)] \tag{5.4}$$

where $N$ is the number of emitter fingers, $J_{BL,E}$ is the leakage hole current density (in A/cm) from the base to emitter, $J_{CL,E}$ is the leakage electron current density (in A/cm) from the emitter to base, $P_E$ is the emitter perimeter length [$P_E = 2(W_E + L_E)$, where $W_E$ and $L_E$ are the emitter finger width and length, respectively (see Figure 5.2)], $F_L$ is an empirical and constant parameter, $V_T = kT/q$ is the thermal voltage, and $J'_{BL,E}$ and $J'_{CL,E}$ are $J_{BL,E}$ and $J_{CL,E}$ for $V_{BE} \gg V_T$ (e.g., fully activated leakage current densities), respectively. The values of $J'_{BL,E}$, $J'_{CL,E}$, and $F_L$ depend on the process. Equations (5.3) and (5.4) suggest that the leakage currents taking place at the emitter periphery increase very rapidly and

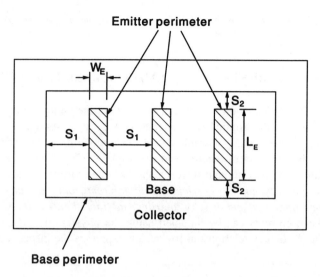

**Figure 5.2** The top view of a three-finger HBT showing the emitter and base perimeters.

become nearly constant as $V_{BE}$ is increased. This trend will later be verified with experimental results.

## 5.2 LEAKAGE CURRENTS AT THE BASE-COLLECTOR PERIPHERY

Since the base-collector junction is reverse biased under forward-active operation, a large electric field exists in the nitride near the base and collector peripheries. This high electric field gives rise to a large probability for the electrons in the p-type GaAs and holes in the n-type GaAs to tunnel through the GaAs-polyimide potential barrier and to reach the collector and base regions, respectively. Similar to the electrons and holes generated from avalanche multiplication in the base-collector depletion region [7], *the electrons entering the collector increase the collector current, whereas the holes entering the base are forced to flow out of the base terminal, which constitutes a negative current component and reduces the total base current.*

Following the same approach as that in the previous section, the base and collector leakage currents occurring at the base perimeter ($I_{BL,B}$ and $I_{CL,B}$) can be modeled as

$$I_{BL,B} = P_B J_{BL,B} = P_B J'_{BL,B}[1 - \exp(-V_{CB}F_L/V_T)] \quad (5.5)$$

$$I_{CL,B} = P_B J_{CL,B} = P_B J'_{CL,B}[1 - \exp(-V_{CB}F_L/V_T)] \quad (5.6)$$

where $J_{BL,B}$ is the leakage hole current density (in A/cm) from the collector to base, $J_{CL,B}$ is the leakage electron current density (in A/cm) from the base to collector, $P_B$ is the base perimeter length, $V_{CB}$ is the applied collector-base voltage, and $J'_{BL,E}$ and $J'_{CL,E}$ are the fully activated $J_{BL,B}$ and $J_{CL,B}$, respectively. For a multifinger HBT, $P_B$ is a function of the emitter finger geometry as well as the finger spacing (Figure 5.2) and is denoted

$$P_B = 2\{[NW_E + (N + 1)S_1] + (L_E + 2S_2)\} \qquad (5.7)$$

where $S_1$ and $S_2$ are the horizontal and vertical spacings of the emitter finger.

## 5.3 LEAKAGE CURRENTS AT THE COLLECTOR-SUBCOLLECTOR PERIPHERY

When subjected to an applied voltage, the $n^+/n$ periphery, similar to the n/p base-collector periphery, can inject free carriers through the interface of dielectric layer covering the periphery and thus give rise to both base and collector leakage currents. These leakage currents have the same bias-dependence as $I_{BL,B}$ and $I_{CL,B}$. However, since the voltage drop of the applied base-collector voltage occurs primarily at the p/n base-collector junction, the voltage drop at the $n^+/n$ high-low junction is negligibly small [8]. As a result, the leakage current at the $n^+/n$ periphery, which is proportional to $[\exp(V'/V_T) - 1]$ ($V'$ is the voltage drop across the high-low junction), can be neglected.

## 5.4 LEAKAGE CURRENTS AT THE SUBCOLLECTOR-SUBSTRATE INTERFACE

Another leakage current is the leakage of free carriers through the $n^+$-GaAs/SI-GaAs (SI denotes semi-insulating) interface. Such an interface is "leaky" due to the very high defect density at the SI GaAs surface, and kinetic factors such as the chemical barrier and diffusion rate often prevent the system from reaching the equilibrium state [9]. Since the SI substrate has a lower electron density and higher hole density than the $n^+$ subcollector, electrons and holes can leak, or diffuse, through the *nonequilibrium* interface and enter the SI substrate and the subcollector, respectively. Because of the high defect density in the SI substrate, electrons entering the SI substrate will recombine with holes via the defect centers. To maintain charge neutrality in the $n^+$ subcollector, electrons are then supplied into the subcollector from the collector contact, which constitutes a current flow $I_{CL,SI}$ *opposite* that of the normal collector current flow. Thus, such a leakage mechanism can be represented by including a negative collector current component $(-I_{CL,SI})$ in the collector current model.

The effects of the $n^+$-GaAs/SI-GaAs leakage on the base current can be more conveniently treated by focusing on the hole transport. Once entering the $n^+$ region, the holes, which are minority carriers, will diffuse across the $n^+$ and n regions. Depending on the recombination process in these regions, a percentage of the holes will reach the base-collector depletion region and be swept into the base region by the large electric field at the base-collector junction. Since the number of holes allowed to be injected into the emitter is fixed by the emitter-base voltage, these extra holes are forced to flow out of the base terminal, which constitutes a current flow *opposite* that of the normal base current flow. Thus, a negative base current component ($-I_{BL,SI}$) also needs to be included in the base current model to account for the n$^+$-GaAs/SI-GaAs interface leakage. Obviously, to maintain quasi neutrality in the SI substrate, holes need to be supplied from a current path through the grounded substrate.

Unlike the leakage process at the emitter-base and base-collector peripheries that require the free carriers to surmount or to tunnel through the potential barrier associated with the dielectric layer, the leakage at the subcollector/substrate interface is caused by the diffusion of electrons and holes through the leaky interface. As a result, the current-voltage relationship used previously is not applicable here. From measured data, we found that $I_{CL,SI}$ and $I_{BL,SI}$ are insensitive to the bias conditions.

## 5.5 BASE AND COLLECTOR CURRENTS INCLUDING BOTH NORMAL AND LEAKAGE COMPONENTS

The normal base current $I_{BN}$ consists of (1) the recombination current $I_{SCR}$ in the emitter-base space-charge region, (2) the surface recombination current $I_{RS}$ at the emitter sidewalls and extrinsic base surface, (3) the recombination current $I_{RB}$ in the quasi-neutral base region, and (4) the injection current $I_{RE}$ from the base into emitter. Thus

$$I_{BN} = I_{SCR} + I_{RS} + I_{RB} + I_{RE}$$

$$= I_1 \exp(V_{BE}/2V_T) + I_2 \exp(V_{BE}/V_T) + I_n(X_2)(1-\alpha) + I_p(X_1)\exp(-V_B/V_T) \quad (5.8)$$

where $I_1$ and $I_2$ are the pre-exponential currents for $I_{SCR}$ and $I_{RS}$, respectively; $I_n(X_2)$ is the electron current at the edge of the quasi-neutral base; $I_p(X_1)$ is the hole current at the edge of the quasi-neutral emitter; $\alpha$ is the base transport factor; and $V_B$ is the valence-band barrier potential across the emitter-base junction. Note that $I_1$ is determined by the Shockley-Read-Hall recombination process, $I_2$ is a function of the surface states and location of Fermi-level pinning, a is influenced by the carrier mobility and lifetime in the base, and $I_n(X_2)$ and $I_p(X_1)$ depend on the emitter and base doping concentrations, layer thicknesses, as well as the heterojunction properties. It is important to point out that, unlike all other base current components that are directly proportional to the emitter area, $I_2$ is proportional to the surface area and thus does not scale with the emitter area [10].

The normal collector current $I_{CN}$ is given by

$$I_{CN} = I_n(X_2)\alpha \quad (5.9)$$

Note that $I_{CN}$ is affected by the carrier transport across the heterojunction, which for an abrupt heterojunction is governed by the thermionic-tunneling-diffusion mechanism. The drift-diffusion theory becomes applicable if the conduction-band spike is effectively removed by grading the Al composition near the heterointerface.

From the preceding discussions, the total base and collector currents ($I_B$ and $I_C$) model including the leakage currents at the emitter-base periphery, base-collector periphery, and n$^+$-GaAs/SI-GaAs interface are

$$I_B = I_{BN} + I_{BL,E} - I_{BL,B} - I_{BL,SI} \quad (5.10)$$

$$I_C = I_{CN} + I_{CL,E} - I_{CL,B} - I_{CL,SI} \quad (5.11)$$

To support the foregoing phenomenological reasoning and modeling, we consider five AlGaAs/GaAs HBTs (HBT-1 to HBT-5) manufactured from different companies/laboratories and thus different processing. The devices have very similar intrinsic make-ups (i.e., doping concentration and layer thickness) but different extrinsic make-ups (i.e., finger pattern, perimeter, and dielectric layer). Table 5.1 gives the detailed finger pattern including the finger number, shape, spacing and the emitter and base perimeter lengths and type of dielectric layer covering the HBT surface.

Figure 5.3(a) shows the Gummel plots of HBT-1 calculated from the present model and obtained from measurements at $V_{CB} = 0$. For this device, empirical parameters $J'_{BL,E} = 8.3 \times 10^{-6}$ A/cm, $J'_{CL,E} = 1.1 \times 10^{-5}$ A/cm, $F_L = 0.005$, and $I_{BL,SI} = I_{CL,SI} = 0$ are used in calculations to give the observed leakage current versus $V_{BE}$ characteristics. The different $J'_{BL,E}$ and $J'_{CL,E}$ may arise from the asymmetrical properties of AlGaAs-polyimide and GaAs-polyimide interfaces at the emitter perimeter. The fact that using $I_{BL,SI} = I_{CL,SI} = 0$ fits the data well indicates that such a device has a good (non-leaky) n$^+$-GaAs/SI-GaAs interface. Note that $I_{BL,B} = I_{CL,B} = 0$ for this case because $V_{CB} = 0$ (see (5.5) and (5.6)), which leads to $I_B \approx I_C$ at small $V_{BE}$.

The relative importance of the base current components at $V_{CB} = 0$ is illustrated in Figure 5.3(b). Clearly, the base leakage current $I_{BL,E}$ is the dominant component for

Table 5.1
HBT Emitter Finger Pattern and Geometry

| Device* | # of Finger | Finger Shape | Finger Area ($\mu m^2$) | Spacing ($\mu m$) | $N \times P_E$ ($\mu m$) | $P_B$ ($\mu m$) | Dieletric Layer |
|---|---|---|---|---|---|---|---|
| HBT-1 | 5 × 5 | Circular | 7 | 10 | 235 | 300 | Polyimide |
| HBT-2 | 6 | Rectangular | 25 | 10 | 150 | 230 | Nitride |
| HBT-3 | 6 | Rectangular | 25 | 10 | 150 | 230 | Nitride |
| HBT-4 | 3 | Rectangular | 40 | 10 | 132 | 172 | Nitride |
| HBT-5 | 4 | Rectangular | 50 | 10 | 180 | 200 | Nitride |

*All HBTs have very similar intrinsic device make-ups such as the doping concentrations and layer thicknesses.

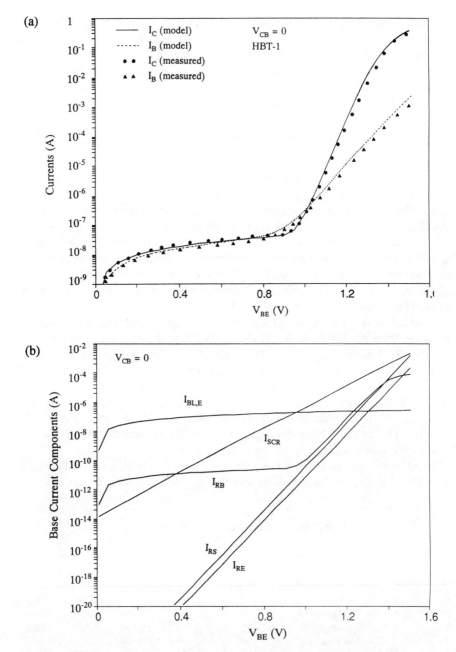

**Figure 5.3** (a) Base and collector currents calculated from the present model and obtained from measurements for HBT-1 at $V_{CB} = 0$ and (b) the corresponding base current components.

$V_{BE} < 0.8$V. For small $V_{BE}$, $I_{RB}$ has the same shape as $I_{BL,E}$ because $I_{RB}$ is proportional to $I_C$ (equation (5.8)), which in turn is equal to $I_{CL,E}$ ($\approx I_{BL,E}$) under such bias conditions. As $V_{BE}$ increases, $I_{BL,E}$ becomes negligibly small and $I_{RB}$ returns to its $V_T$-like slope. The surface recombination current $I_{RS}$, like $I_{RE}$, also possesses a $V_T$-like slope [10] and becomes more important as $V_{BE}$ is increased. On the other hand, $I_{SCR}$ has a $2V_T$-like slope due to the dominance of Shockley-Read-Hall recombination process in the space-charge region.

Figure 5.4(a) shows the Gummel plots of HBT-1 at $V_{CE} = 2.5$V. Note that $V_{CB}$ (= $V_{CE} - V_{BE}$) varies with $V_{BE}$ in this figure. In addition to the leakage current densities at the emitter perimeter mentioned earlier, $J'_{BL,B} = 7 \times 10^{-6}$ A/cm and $J'_{CL,B} = 1.4 \times 10^{-5}$ A/cm are also used here to describe the base perimeter leakage. It should be pointed out that assuming $J'_{BL,B} = J'_{CL,B}$ in calculations will result in $I_B = I_C$ at $V_{BE} = 0$ for all $V_{CE}$ (see (5.10) and (5.11)), a trend not supported by measurement data (see Figure 5.4(a)). Under the bias condition considered in Figure 5.4(a), the leakage current at the base perimeter is larger than that at the emitter perimeter ($I_{BL,B} > I_{BL,E}$ and $I_{CL,B} > I_{CL,E}$ in (5.10) and (5.11)). This is evidenced in the results shown in Figure 5.4(b). For $V_{CE} > 0$ and small $V_{BE}$, the larger leakage at the base perimeter than emitter perimeter, together with the negligibly small $I_{BN}$ and $I_{CN}$, yields the observed *negative* base current (current flow is in the direction opposite to that of the normal flow) and near constant collector current.

Figures 5.5(a, b) show the $I_C$ and $I_B$ characteristics calculated from the model and obtained from measurements of HBT-2 at $V_{CB} = 0$ and $V_{CE} = 2.5$V, respectively. Constant values of $I_{BL,SI} = I_{CL,SI} = 5 \times 10^{-8}$ A are used in calculations, suggesting that HBT-2 has a leaky $n^+$-GaAs/SI-GaAs interface. This leakage leads to a trend differs considerably from that seen in HBT-1. Both $I_B$ and $I_C$ have a dip at $V_{BE} \approx 1$V and $V_{CB} = 0$ ($I_B$ and $I_C$ are negative for $V_{BE} < 1$V). At $V_{CE} = 2.5$V, however, only $I_B$ has a dip, and $I_C$ is positive for all $V_{BE}$. It is equally important to note that while HBT-1 and HBT-2 have different types of dielectric layer, they nonetheless have very similar leak current densities at the emitter and base peripheries (see Table 5.2). This suggests that the leakage mechanism at the peripheries is insensitive to the type of passivation layer but rather is strongly affected by the property of the interface, which varies from process to process. Conversely, it is likely that HBTs that have the same type of dielectric layer can have considerable different leakage current characteristics. This point is evidenced by the leakage current densities of HBT-2 to HBT-5 (all have nitride layers) given in Table 5.2.

We now turn our attention to another HBT (HBT-3). This device has the same finger pattern and geometry as HBT-2 but was fabricated from a different lot. Compared to HBT-2, HBT-3 has a very similar $I_C$ behavior but very different $I_B$ characteristics, as shown in Figures 5.6(a, b). The present model can still be used to describe such $I_B$ and $I_C$ characteristics, provided $I_{BL,SI}$ and $I_{BL,B}$ are neglected in model calculations. The large $I_{CL,SI}$ and zero, or very small, $I_{BL,SI}$ required in calculations may arise from the fact that this HBT has a leaky subcollector/substrate interface as well as a high electron-hole recombination process in collector and subcollector. As a result, the vast majority of holes that leak from the substrate into the subcollector are recombined with electrons in the $n^+$

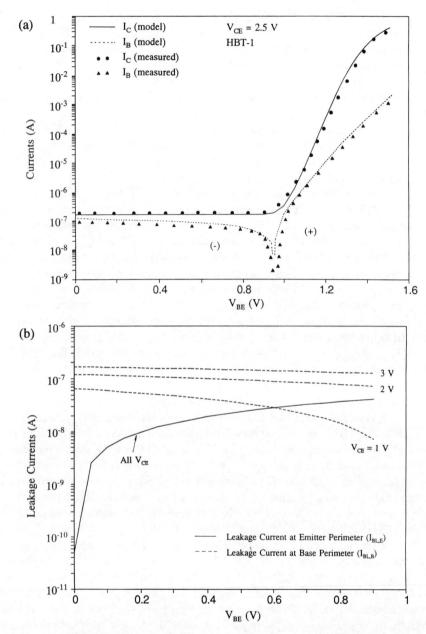

**Figure 5.4** (a) Base and collector currents calculated from the present model and obtained from measurements for HBT-1 at $V_{CE} = 2.5$ V and (b) the leakage currents versus $V_{BE}$ calculated for three different $V_{CE}$.

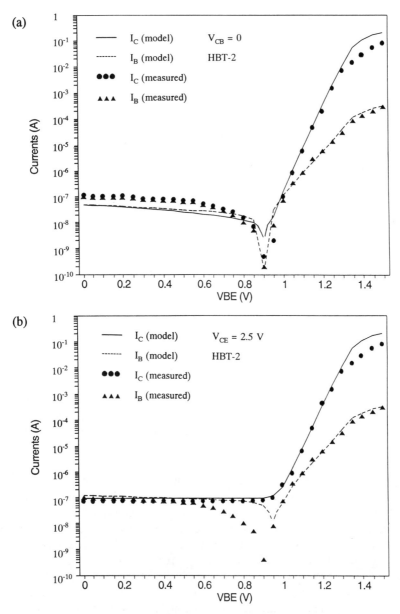

**Figure 5.5** Base and collector currents calculated from the present model and obtained from measurements for HBT-2 at (a) $V_{CB} = 0$ and (b) $V_{CE} = 2.5$V.

**Table 5.2**
Leakage Current Components Used in Calculations

| Device | $J'_{BL,E}$ (A/cm) | $J'_{CL,E}$ (A/cm) | $J'_{BL,B}$ (A/cm) | $J'_{CL,B}$ (A/cm) | $I_{BL,SI}$ (A) | $I_{CL,SI}$ (A) |
|---|---|---|---|---|---|---|
| HBT-1 | $8.3 \times 10^{-6}$ | $1.1 \times 10^{-5}$ | $7.0 \times 10^{-6}$ | $1.4 \times 10^{-5}$ | 0.0 | 0.0 |
| HBT-2 | $1.3 \times 10^{-5}$ | $1.7 \times 10^{-5}$ | $9.0 \times 10^{-6}$ | $1.8 \times 10^{-5}$ | $5.0 \times 10^{-8}$ | $5.0 \times 10^{-8}$ |
| HBT-3 | $1.3 \times 10^{-5}$ | $1.7 \times 10^{-5}$ | 0.0 | 0.0 | 0.0 | $5.0 \times 10^{-8}$ |
| HBT-4 | $5.0 \times 10^{-4}$ | $4.0 \times 10^{-4}$ | $1.0 \times 10^{-4}$ | $5.0 \times 10^{-4}$ | $5.0 \times 10^{-8}$ | $5.0 \times 10^{-8}$ |
| HBT-5 | $3.5 \times 10^{-4}$ | $7.0 \times 10^{-4}$ | $4.2 \times 10^{-4}$ | $1.1 \times 10^{-3}$ | 0.0 | 0.0 |

HBT-1: Average emitter-base periphery, average base-collector periphery, good subcollector/substrate interface.
HBT-2: Average emitter-base periphery, average base-collector periphery, leaky subcollector/substrate interface.
HBT-3: Average emitter-base periphery, good base-collector periphery, leaky subcollector/substrate interface and substantial electron-hole recombination in the collector and subcollector.
HBT-4: Poor emitter-base periphery, poor base-collector periphery, leaky subcollector/substrate interface.
HBT-5: Poor emitter-base periphery, poor base-collector periphery, good subcollector/substrate interface.

and n regions before they reach the base region. On the other hand, the negligible $I_{BL,B}$ may stem from a good quality base-collector perimeter of the HBT, thus minimizing the leakage at the periphery. The different trends obtained from the identical HBT-2 and HBT-3 illustrate that the HBT leakage currents are highly process dependent.

Two more HBTs (HBT-4 and HBT-5) that have finger patterns and geometries similar to the HBT-3 but are fabricated from different manufactures are also investigated, and their leakage behaviors are successfully described by the model, as shown in Figures 5.7(a,b) and 5.8(a,b), respectively. Fitting the model and measurements, we found that both devices have relatively large leakage current densities at the base and base peripheries (see Table 5.2). Their different leakage behaviors shown in Figures 5.7 and 5.8 arise from the fact that the subcollector/substrate interface leakage is minimal in HBT-5 and is large in HBT-4.

From the foregoing analysis, it can be seen that the HBT leakage currents are affected by the six current components in a complex kind of manner, particularly for HBTs exhibiting negative base and/or collector currents (i.e., HBT-2, HBT-3, and HBT-4). Here, we summarize the modeling procedure to make the determination of leakage current components more efficient. Since there is only one negative current component (the leakage current $I_{CL,SI}$ due to the leaky n$^+$ GaAs/SI GaAs interface) in the total collector current (refer to (5.11)), the negative collector leakage current can be attributed to $I_{CL,SI}$. The positive portion of the collector leakage current is then fitted by the two leakage current components ($I_{CL,E}$ and $I_{CL,B}$) that occurred at the emitter and base perimeters. The base leakage current, on the other hand, is more complicated than its collector counterpart because it involves two negative ($I_{BL,B}$ and $I_{BL,SI}$) and one positive components ($I_{BL,E}$) (see (5.10)). When a negative base current is observed in measurement at $V_{CB} = 0$, $I_{BL,SI}$ is used in the model to give a best fit to the experimental data. The other negative component $I_{BL,B}$, which is a function of $V_{CB}$, is then used to fit the

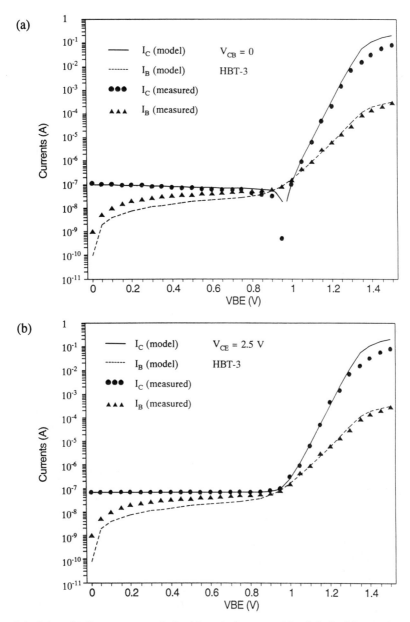

**Figure 5.6** Base and collector currents calculated from the present model and obtained from measurements for HBT-3 at (a) $V_{CB} = 0$ and (b) $V_{CE} = 2.5$V.

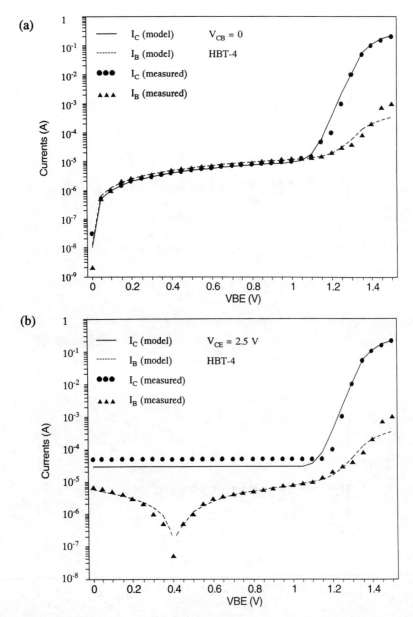

**Figure 5.7** Base and collector currents calculated from the present model and obtained from measurements for HBT-4 at (a) $V_{CB} = 0$ and (b) $V_{CE} = 2.5$V.

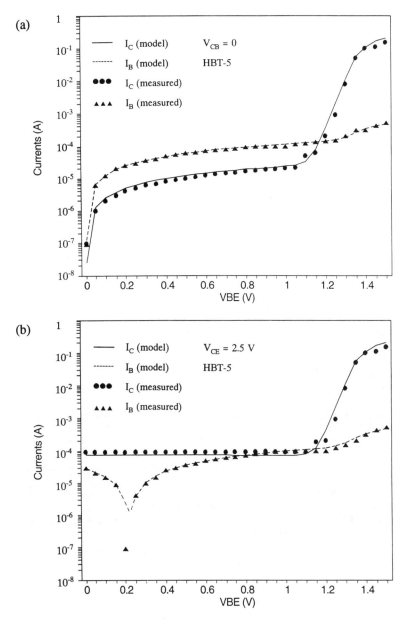

**Figure 5.8** Base and collector currents calculated from the present model and obtained from measurements for HBT-5 at (a) $V_{CB} = 0$ and (b) $V_{CE} = 2.5$V.

negative base current at $V_{CB} > 0$ (i.e., $V_{CE} = 2.5V$). The positive portion of the base leakage current is fitted using $I_{BL,E}$. If negative base and collector leakage currents are absent at $V_{CB} = 0$, then $I_{BL,SI} = I_{CL,SI} = 0$, as for the case of HBT-1 and HBT-5.

Table 5.2 summarizes the leakage current components used in model calculations to fit the experimental data. Based on these values, comments on the property of HBTs' emitter-base periphery, base-collector periphery, and subcollector-substrate interface are also given (see Table 5.2).

The above study has shown that their is a strong correlation between the HBT leakage currents, which are related to the HBT surface quality, and long-term performance. In general, the leakage currents serve as an indicator of the HBT reliability; the larger the leakage currents, the larger the HBT current gain long-term degradation.

## References

[1] Parikh, C. D., and F. A. Lindholm, "Space-Charge Region Recombination in Heterojunction Bipolar Transistors," *IEEE Trans. Electron Devices*, Vol. 39, 1992, p. 2197.
[2] Sandroff, C. J., R. N. Nottengurg, J. C. Bischoff, and R. Bhat, "Dramatic Enhancement in the Gain of GaAs/AlGaAs Heterostructure Bipolar Transistor by Surface Chemical Passivation," *Appl. Phys. Lett.*, Vol. 51, 1987, p. 33.
[3] Tiwari, S., D. J. Frank, and S. L. Wright, "Surface Recombination in GaAlAs/GaAs Heterojunction Bipolar Transistors," *J. Appl. Phys.*, Vol. 64, 1988, pp. 5009–5012.
[4] Liu, W., D. Costa, and J. S. Harris, "Theoretical Comparison of Base Bulk Recombination Current and Surface Recombination Current of a Mesa AlGaAs/GaAs Heterojunction Bipolar Transistor," *Solid-St. Electron.*, Vol. 34, 1991, p. 119.
[5] Aspnes, D. E., "Recombination at Semiconductor Surfaces and Interfaces," *Surface Sci.*, Vol. 132, 1983, p. 406.
[6] Ash, M. S., and H. C. Gorton, "A Practical End-of-Life Model for Semiconductor Devices," *IEEE Trans. Reliability*, Vol. 38, 1989, p. 485.
[7] Liou, J. J., and J. S. Yuan, "Modeling the Reverse Base Current Phenomenon Due to Avalanche Effect in Advanced Bipolar Transistors," *IEEE Trans. Electron Devices*, Vol. 37, 1990, p. 2274.
[8] Warner, Jr., R. M., and B. L. Grung, *Transistors: Fundamentals for the Integrated-Circuit Engineer*, New York: Wiley, 1983, Chapter 6.
[9] Wilmsen, C. M., "Insulator/Semiconductor Interfaces," in *The Physics of Submicron Semiconductor Devices*, edited by H. L. Grubin et al., New York: Plenum Press, 1988.
[10] Liu, W., and J. S. Harris, Jr., "Diode Ideality Factor for Surface Recombination Current in AlGaAs/GaAs Heterojunction Bipolar Transistors," *IEEE Trans. Electron Devices*, Vol. 39, 1992, p. 2726.

# Chapter 6
# Noise Characteristics of HBTs

Noise is an important issue in semiconductor device design since it sets a lower limit to the strength of signals that can be processed electronically [1–2]. In this chapter, we first give an overview of the noise characteristics of the HBT. The minimum noise factor, a noise figure of merit for HBT operating at high frequency, is then treated in detail.

## 6.1 OVERVIEW

For junction diodes and bipolar transistors, the conventional noise model [1–3] suggests the existence of three distinct regions in the noise spectra: a $1/f$ shape (Flicker or $1/f$ noise) at lower frequencies, a Lorenztian spectrum (bump or burst noise) at intermediate frequencies, and a constant noise (white noise or shot noise) at higher frequencies. Figure 6.1 gives the equivalent base current noise density ($A^2/Hz$) for three different collector current levels that show such characteristics.

Low-frequency noise is related to the extrinsic base surface property [3,4], and most researchers suggest that the $1/f$ noise originated from fluctuations in the occupancy of surface traps, which in turn perturbs the extrinsic base surface recombination current $I_{bs}$ [5]. As a result, the magnitude of $1/f$ noise serves as an indicator for the recombination mechanism at the surface. Assuming the initial surface recombination velocity $S$ ($S \approx 5 \times 10^5$ cm/s without an AlGaAs ledge structure on the base surface and $S \approx 10^4$ cm/s with a ledge) is independent of position and excess free-carrier density at the surface, Fonger [6] showed that the initial base current $1/f$ noise spectral density $S_{Ib}$ can be expressed as

$$S_{Ib}(f) = (I_{bs}^2/S^2)S_s(f) \qquad (6.1)$$

where $S_s(f)$ is the spectral density characterizes the noise contribution at the surface and is proportional to $1/f$. If we further assume that

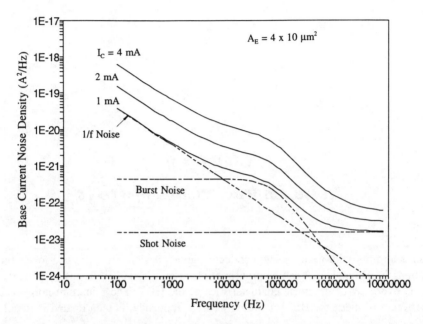

**Figure 6.1** Equivalent base current noise density versus frequency for three different current levels, with three noise components identified.

$$I_{bs} \approx qSn(0)L_d P_E \tag{6.2}$$

and that $n(0)$ can be expressed in terms of $I_C$ because, using the approximation that the electron-hole recombination in the quasi-neutral base is absent,

$$I_C \approx qA_E D_n n(0)/W_B \tag{6.3}$$

Here $L_D$ is the electron lateral diffusion length in the base, $n(0)$ is the injected minority electron carrier at the base edge of the space-charge region, and $P_E$ is the emitter perimeter length. Putting (6.2), (6.3) into (6.1) yields

$$S_{Ib}(f) \approx I_C^2 (W_b L_d / D_n)^2 (P_E / A_E)^2 S_s(f) \tag{6.4}$$

Burst noise in bipolar transistors is often associated with traps or generation-recombination (g-r) centers near the emitter-base space-charge region. It could be attributed to the repeated switching between two g-r centers, at intervals of the order of a millisecond [7]. Previous phenomenological studies showed that the spectral density of burst noise takes the form of a Lorenztian [3]:

$$S_{\text{Iburst}}(f) \propto \tau I_C^m / [1 + (2\pi\tau f)^2] \tag{6.5}$$

where $\tau$ is the time constant of trap involved and $m$ is an empirical constant in the range of 0.5 and 2. The time constant is thermally activated by an energy $E_a$:

$$\tau = (\tau_0/T^2)\exp(E_a/kT) \tag{6.6}$$

The Lorenztian spectrum has a large temperature dependence, as shown in Figure 6.2, because the time constant decreases with increasing temperature as predicted by (6.6).

The concept of shot noise originated with the vacuum diode [7]. In the absence of a space charge an electron charge $q$, which was emitted from the cathode, traveled without hindrance to the anode. This electron transit constituted a current pulse. The essential ideas were that the emissions were random in time and that both emissions and transits were uncorrelated. The application of the above theory to solid-state devices is limited, however, owing to the absence of uninterrupted transits between electrodes. Shot noise in a junction device can be visualized as arising from the fact that the current flow consists of individual pulses with random spacing. If transits are uncorrelated, the resulting fluctuations of current will appear as shot noise. A characteristic of shot noise is that its spectrum is white (i.e., constant) over frequencies for which the transit time is relatively small. For these frequencies, van der Ziel [8] showed that the shot noise associated with a dc current $I$ is equal to $2qI$. This model often fails at very high frequencies (i.e., > 1 GHz), however, as the noise figure at such frequencies exhibits nonuniform characteristics versus frequency [9].

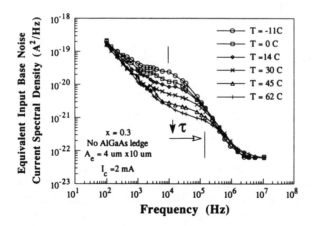

**Figure 6.2** Temperature dependence of the burst noise (Lorenztian spectrum) measured at different temperatures (*Source*: [3]. © 1992 IEEE.).

## 6.2 HIGH-FREQUENCY NOISE CHARACTERISTICS OF HBTs

For AlGaAs/GaAs HBTs used in microwave applications, shot noise is of major concern [10, 11]. Attempts have been made in the past to account for the frequency dependence in shot noise [9–13]. For example, Hawkin [12] derived such a model based on the small-signal equivalent circuit and the concept that the charge transport is modulated by the emitter junction capacitance at high frequencies. However, the model neglects other bias-dependent elements such as the diffusion capacitance and the base-collector junction capacitance. A more physical approach has been suggested to explain the frequency-dependent shot noise in [13]. It is established from the theory that there is a delay time associated with the current transport in the device and that the effects of such a delay time must be included in the shot nose model at high frequencies. Both models also account for the bias current dependence, but they assume that the HBT high-frequency noise is independent of the applied voltage (i.e., collector-emitter voltage $V_{CE}$)—an approximation that contradicts the experimental results. In addition, the effects of the emitter area of the HBT on the noise performance have been overlooked in the literature.

This section characterizes the AlGaAs/GaAs HBT noise at high frequencies and presents a physics-based model to describe the shot noise behavior in the HBT. The model extends the approach in [13] and can satisfactorily predict the current, voltage, and emitter area dependencies of noise characteristics observed in measurements. Furthermore, the frequency-dependent nature of noise at very high frequencies is incorporated in the model. The following sections describe the HBT noise model, the self-heating effect in the HBT [14,15] that gives rise to the voltage dependence of shot noise at high frequencies, and comparisons of calculated noise characteristics with measurements.

### 6.2.1 Noise Model

Neglecting the feedback and assuming a resistive signal source, the equivalent noise voltage spectral density $e_N^2$ (V$^2$/Hz) in a bipolar device is given by [2]

$$e_N^2 = 4kTr_{bb} + 2qI_B r^2_{bb} + 2qI_C(r_{bb} + r_{be})^2 g_m^{-2} r_{be}^{-2}$$
$$+ K_F r^2_{bb} I_B^2/f + K_B r_{be}^2 I_B^2 \tau/[1 + (\omega\tau)^2] \quad (6.7)$$

where the first term on the right-hand side of (6.7) is the thermal noise in the base resistance $r_{bb}$, the second term is the shot noise in the base current, the third term is the shot noise in the collector current, the fourth term is the Flicker (1/$f$) noise in the base current, and the last term is the burst noise (Lorenztian spectrum) caused by the random telegraph signal arising from switching between emission and capture states at recombination centers. In general, 1/$f$ noise dominates in low frequencies, burst noise dominates in medium frequencies, and shot noise dominates in high frequencies [3]. Since the high-frequency noise characteristics are of interest in this study, the Flicker noise will not be emphasized.

In (6.7), $T$ is the lattice temperature of the HBT, which can be higher than the ambient temperature due to the self-heating effect (discussed in chapter 4); $I_B$ is the base current; $I_C$ is the collector current; $g_m = (q/kT)I_C$ is the transconductance; $r_{be} = \beta/gm$ ($\beta$ is the dc current gain) is the small-signal base-emitter resistance; $K_F$ is the Flicker coefficient, which is proportional to the recombination velocity at the base surface; $f$ is the frequency; $K_B$ is the Lorenztian coefficient associated with the burst noise; w is the radian frequency; and t is the effective free-carrier lifetime at a trapping center [3] as is expressed as

$$\tau = (\tau_0/T^2)\exp(E_a/kT) \tag{6.8}$$

where $\tau_0$ is a constant and $E_a$ is the activation energy of the trapping state. Clearly, $\tau$, and thus the Lorenztian spectrum governing the burst noise, is highly temperature dependent. As will be shown later, such a temperature dependence influences considerably the medium- and high-frequency noise characteristics in the HBT.

The conventional shot noise expressions given in (6.7) are valid only up to frequencies comparable to the inverse of the delay time associated with the free-carrier transport in the device. Therefore, the shot noise $e^2_{\text{Ib,shot}}$ in the base current must be corrected at high frequencies, where the delay time associated with the hole charge passing through the n-type emitter needs to be addressed. The base current shot noise can be estimated as [13]

$$e^2_{\text{Ib,shot}} = 2qI_B r^2_{bb}(2G_{Ib} - 1) \tag{6.9}$$

where

$$G_{Ib} = \text{Re}[(2j\omega\tau_{db})^{0.5}/\tanh(2j\omega\tau_{db})^{0.5}] \tag{6.10}$$

and $\tau_{db}$ is the delay time associated with the base current flow [16] and is denoted

$$\tau_{db} = W_E^2/2D_p + (kT/q)C_{jE}/I_C \tag{6.11}$$

$W_E$ is the emitter thickness, $D_p$ is the hole diffusion coefficient in the emitter, and $C_{jE}$ is the emitter-base junction capacitance. The first term on the right-hand side of (6.11) is the emitter transit time, and the second term is the emitter capacitance charging time.

Likewise, the shot noise $e^2_{\text{Ic,shot}}$ in the collector current is

$$e^2_{\text{Ic,shot}} = 2qI_C(r_{bb} + r_{be})^2 g_m^{-2} r_{be}^{-2}(2G_{Ic} - 1) \tag{6.12}$$

where

$$G_{Ic} = \text{Re}[(2j\omega\tau_{dc})^{0.5}/\tanh(2j\omega\tau_{dc})^{0.5}] \quad (6.13)$$

and $\tau_{dc}$ is the delay time associated with the collector current flow [16] and is denoted

$$\tau_{dc} = X_{scr}/2v_d + R_c C_{jC} + W_B^2/2D_n \quad (6.14)$$

$X_{scr}$ is the base-collector space-charge layer thickness, $v_d$ is the drift velocity, $R_C$ is the collector resistance, $C_{jC}$ is the base-collector junction capacitance, $W_B$ is the base thickness, and $D_n$ is the electron diffusion coefficient. The three terms on the right-hand side of (6.14), in the same order as in (6.14), are the collector signal delay time, the collector charging time, and the base transit time, respectively.

Similarly, the equivalent shot noise current spectral density $i_N^2$ (A²/Hz) can be derived as [2]

$$i_N^2 = 2qI_B(2G_{Ib} - 1) + 2qI_C g_m^{-2} r_{be}^{-2}(2G_{Ic} - 1) + K_F I_B^2/f + K_B I_B^2 \tau/[1 + (\omega\tau)^2] \quad (6.15)$$

An index of noise performance is the noise factor $F$, which is defined as the ratio of the total noise output power to the noise output power arising from thermal noise in the source resistance alone. However, a figure of merit commonly used to describe the HBT noise performance at high frequencies is the minimum noise factor $F_{min}$, which is obtained by optimizing the noise factor with respect to the source resistance and is found to be related to the noise power $(e_N i_N)$ as [2]

$$F_{min} = 1 + e_N i_N/2kT \quad (6.16)$$

For a typical HBT, our calculations show that the base current shot noise is larger than the collector current shot noise. Hence the frequency dependence of $F_{min}$ is influenced mainly by $\tau_{db}$. This result seems to justify the approximation made by Hawkin [12] that the effects of the diffusion capacitance and base-collector junction capacitance on the noise factor are negligible.

### 6.2.2 Thermal Effect on Noise Behavior

From the foregoing analysis, it is clear that the lattice temperature $T$ in the HBT strongly affects the noise spectral density and $F_{min}$. An analytical model was developed in Chapter 4 to correlate the HBT current and the lattice temperature. We now briefly review the model.

The heat power $P_s$ generated in the HBT is

$$P_s = J_C V_{CE} A_E \tag{6.17}$$

where $A_E$ is the emitter area, $J_C$ is the collector current density, and $V_{CE}$ is the applied collector-emitter voltage. The generated heat power is related to the lattice temperature as

$$T - T_0 = P_s R_{th} \tag{6.18}$$

where $T_0 = 300K$ is the ambient temperature and $R_{th}$ is the thermal resistance of the semi-insulating GaAs substrate. Assuming that the heat is dissipated primarily through the effective area $A_{eff}$ in the substrate with a lateral diffusion angle $\theta$ ($\theta = 45$ deg is used [17]), together with the Kirchhoff transformation [18], yields the temperature-dependent $R_{th}$, which is denoted

$$R_{th} = (\eta - T_0)/P_s \tag{6.19}$$

where

$$\eta = [1/T_0^{b-1} - (b - 1)R_{th0}P_s/T_0^b]^{-1/b-1} \tag{6.20}$$

$b = 1.22$, and $R_{th0}$ is the thermal resistance for the case that the thermal conductivity in the substrate is assumed to be temperature independent and is denoted [14]

$$R_{th0} = 1/K_{s0} \int_0^{X_s} dx/A_{eff}(x) = 1/K_{s0} \int_0^{X_s} dx/[A_C + 2Z \tan(\theta)x] \tag{6.21}$$

where $K_{s0}$ is the thermal conductivity at $T_0$ ($K_{s0} = 0.47$ W/k-cm at 300K), $Z$ is the HBT width, $A_C$ is the collector area, and $X_s$ is the thickness of the substrate.

For a given $V_{CE}$, the initial value of $T$ can be calculated from the above equations after the initial $J_C$ and $P_s$ are calculated at the ambient temperature. This temperature is then used to calculate the initial $R_{th}$. The correct $T$ and $J_C$ are obtained after several iterations. Figure 6.3 illustrates the HBT current-voltage characteristics of decreasing $I_C$ versus $V_{CE}$ behavior (negative conductance) at high base current levels due to the self-heating effect.

We first consider an N/p/n AlGaAs/GaAs HBT with a typical device make-up [19]. The device (HBT-351) has an emitter area of 35 µm² (5-dot, with 3-µm diameter for each dot, and 1-finger structure, see Figure 6.4(a)), and $r_{bb} = 12$ Ω. The related current gain and current density data for the HBT are given in Table 6.1. Also, $E_a$ was assumed to be 0.2 eV [3] in calculations.

The noise measurement procedure is as follows. The microwave noise parameters were measured from 2 GHz to 18 GHz. The measurement system consisted of an ATN-NP5 wafer probe test set, HP8971C noise figure test set, HP8970B noise figure meter, and HP8510B network analyzer. Such a system permitted on-wafer, small-signal character-

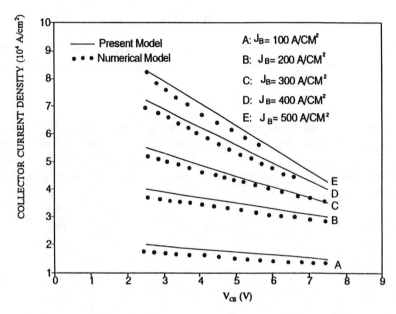

**Figure 6.3** HBT current-voltage characteristics as a function of the base current density.

ization of the devices. The noise figure and associated gain of the device under test were measured at various impedance matching and dc bias conditions. Then the minimum noise factor and the optimal source impedance were extracted from the measured noise figure at a particular dc bias.

Figure 6.5 shows $F_{min}$ versus frequency characteristics for four different base currents calculated from the present model and obtained from measurements. The results indicate that $F_{min}$ increases with $I_B$ and increases with frequency when both frequency and $I_B$ are relatively high. The former increase results because the shot noise is proportional to $I_B$ (see (6.7)), and the latter increase is caused by the fact that the minority-carrier transport delay time becomes significant at high frequencies (see (6.9) to (6.14)). Note that $F_{min}$ would be constant versus frequency if the conventional frequency-independent shot noise model, which assumes zero charge transport delay time, is used. Also, since the collector-emitter voltage $V_{CE}$ considered here is relatively low (e.g., 2V), the self-heating effect is not important. Hence the lattice temperature is found to range from only 300K to 325K for the four different $I_B$ considered.

If $V_{CE}$ is increased and $I_B$ is fixed, the lattice temperature will increase. This will reduce $F_{min}$ because it is approximately proportional to $C_1 + C_2/T$, where $C_1$ and $C_2$ are constants, as evidenced by the results shown in Figure 6.6 calculated and measured for

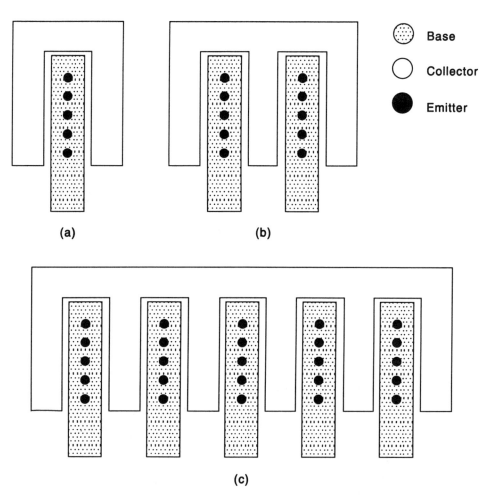

**Figure 6.4** Schematic illustrating the emitter finger pattern of (a) a 5-dot, 1-finger HBT; (b) a 5-dot, 2-finger HBT; and (c) a 5-dot, 5-finger HBT.

three different $V_{CE}$. The same $F_{min}$ versus $V_{CE}$ trend is also found in other HBTs. We want to point out that since the $F_{min}$ changes at different $V_{CE}$ are small (within one- or two-tenths of decibels), they are within the error of margin of measurements. Based on the HBT self-heating model discussed above, the lattice temperatures were estimated to be 315K, 385K, and 445K for $V_{CE}$ = 2V, 3V, and 4V, respectively. Note that the temperature does not increase linearly with $V_{CE}$ but rather is approximately proportional to $V_{CE}^{0.5}$ for the

**Table 6.1**
Current Gain and Current Density of Three HBTs Considered

| Device | $V_{CE}$ (V) | $I_C$ (mA) | $I_B$ (mA) | $J_C$ ($10^5$ A/cm$^2$) | Current Gain |
|---|---|---|---|---|---|
| HBT-351 | 2 | 3.55 | 100 | 0.1 | 35.5 |
| HBT-351 | 3 | 3.53 | 100 | 0.1 | 35.3 |
| HBT-351 | 4 | 3.49 | 100 | 0.1 | 34.9 |
| HBT-352 | 2 | 1.30 | 50 | 0.02 | 26 |
| HBT-352 | 2 | 2.04 | 75 | 0.03 | 27.2 |
| HBT-352 | 2 | 2.59 | 100 | 0.04 | 25.9 |
| HBT-355 | 2 | 1.86 | 75 | 0.01 | 24.8 |
| HBT-355 | 2 | 2.55 | 100 | 0.01 | 25.5 |
| HBT-355 | 2 | 5.46 | 200 | 0.03 | 27.3 |
| HBT-355 | 2 | 33.3 | 1000 | 0.2 | 33.3 |
| HBT-355 | 3 | 68.9 | 2000 | 0.4 | 34.4 |

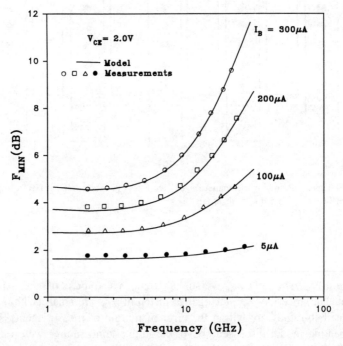

**Figure 6.5** Minimum noise factor calculated from the model and obtained from measurements for a constant voltage and four different base current levels.

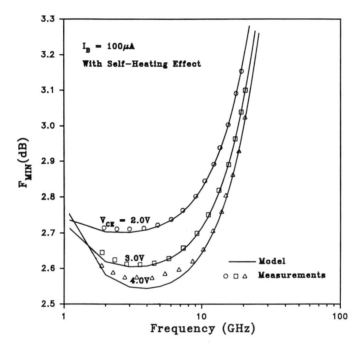

**Figure 6.6** Minimum noise factor calculated from the model and obtained from measurements for a constant base current and three different voltages.

HBT considered because the heat power generated in the HBT is equal to $I_C V_{CE}$, and $I_C$ actually decreases as $V_{CE}$ is increased due to the elevated lattice temperature in the HBT (see Figure 6.3). Consequently, the lattice temperature does not increase in direct proportion with $V_{CE}$.

If the self-heating effect is not accounted for in the model (e.g., $T = 300$K for all $V_{CE}$), then the same $F_{min}$ are predicted for all voltages, which is illustrated in Figure 6.7.

One may speculate that the $F_{min}$ decrease is a consequence of increasing $C_{jC}$ as $V_{CE}$ is increased. This cannot be the case for the following two reasons. First, as mentioned earlier, our calculations show that the shot noise in the collector is not as significant as that in the base current. As a result, changing $C_{jC}$, which changes the delay time in the collector current shot noise, does not affect the overall noise behavior. Second, even if the collector current shot noise were the dominant term, increasing $C_{jC}$ will decrease the onset frequency for which the noise starts to increase but will not change the plateau of noise as observed in the measurements. This point will be addressed in more detail when the effect of the emitter size on the noise characteristics is examined.

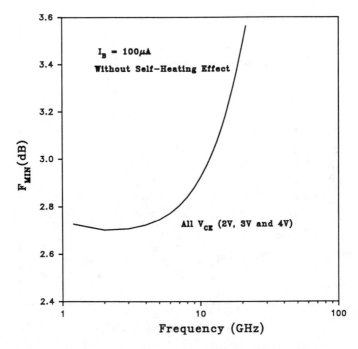

**Figure 6.7** Minimum noise factor calculated from the model without including the self-heating effect.

An interesting phenomenon observed in Figure 6.6 is that $F_{min}$ actually first decreases with frequency and then increases above 5 GHz. As mentioned earlier, the $F_{min}$ increase is due to the nonzero minority-carrier transport delay time in the device, but the decrease cannot be explained by the shot noise behavior alone. Also, the degree of $F_{min}$ decrease increases with $V_{CE}$. To investigate this, the noise current spectral density ($i_N^2$), including all noise components, were calculated versus frequency for different $V_{CE}$, and the results are given in Figure 6.8. It can be seen that increasing $V_{CE}$, and thus increasing the lattice temperature, decreases the plateau and increases the onset frequency at which the burst noise starts to fall. The same trend has been reported in [3], which measured the noise current spectral density versus frequency for various temperatures. Hence this mechanism contributes to the $F_{min}$ decrease versus frequency found in Figure 6.6. Note that the noise spectral density at low frequencies is dominated by the $1/f$ noise (Figure 6.8).

Figure 6.9 depicts both the noise voltage and current spectral densities calculated

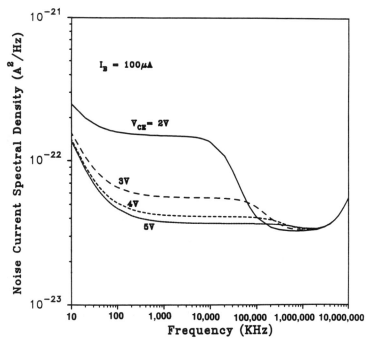

**Figure 6.8** Noise current spectral density calculated from the model for a constant base current and different voltages.

from the present model for three different bias currents. All four distinct regions in the noise spectra are clearly illustrated a $1/f$ shape at lower frequencies, a Lorenztian spectrum at intermediate frequencies, a constant shot noise at higher frequencies, and an increasing shot noise at very high frequencies.

### 6.2.3 Effect of Emitter Area on Noise Behavior

We now investigate the effect of emitter area on the HBT noise performance. In addition to the 5-dot, 1-finger HBT (HBT-351) used previously, two more HBTs (HBT-352 and HBT-355) that have 5-dot, 2-finger and 5-dot, 5-finger emitter patterns are considered (see Figures 2(b, c)). Their current gain and current density information are also listed in Table 6.1. Since the dot sizes in all three HBTs are the same (3-μm diameter), the 5-finger HBT has a larger emitter area (175 μm²) than the 2- and 1-finger HBTs (i.e., 70 μm² and 35 μm²). Figure 6.10 illustrates $F_{min}$ versus the frequencies calculated from the model and obtained from measurements for the three devices at $I_B = 100$ μA and $V_{CE} = 2$V. Since

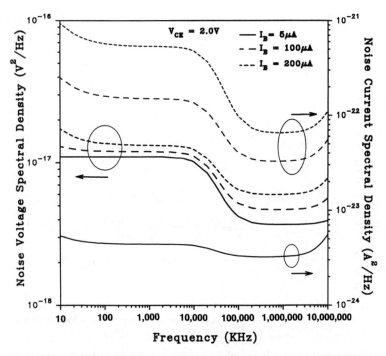

**Figure 6.9** Noise voltage and noise current spectral densities calculated from the model for a constant voltage and three different currents.

$V_{CE}$ is relatively small, the variation of $F_{min}$ found in the figure is not attributed to the self-heating effect. Note the decrease in the plateau of $F_{min}$ as the number of fingers, and thus the emitter area, is increased. Such a reduction in $F_{min}$ is attributed to the decrease in the base resistance, which is inversely proportional to the emitter area [20]. On the other hand, the decrease of the onset frequency at which $F_{min}$ starts to rise results from an increase in the free-carrier transport delay time. As mentioned earlier, the frequency dependence of noise at a high frequency is influenced mainly by the delay time associated with the base current flow, which involves the emitter transit time and the emitter charging time through the emitter-base junction capacitance $C_{jE}$. Since $C_{jE}$ increases with the emitter area, the delay time increases, and the onset frequency for $F_{min}$ decreases with increasing emitter area (Figure 6.10).

The above results show that the HBT high-frequency noise characteristics are strongly affected by minority-carrier delay time and self-heating, and demonstrate the importance of including these effects in modeling the HBT high-frequency noise behavior.

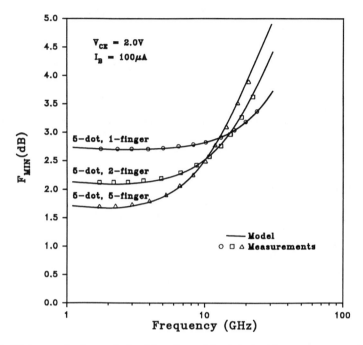

**Figure 6.10** Minimum noise factor calculated from the model and obtained from measurements for three HBTs with different emitter areas at $I_B = 100$ µA and $V_{CE} = 2$V.

## References

[1] van der Ziel, A., *Noise in Solid State Devices and Circuits,* New York: Wiley, 1986.
[2] Fish, P. J., *Electronic Noise and Low Noise Design,* New York: McGraw-Hill, 1994.
[3] Costa, D., and J. S. Harris, Jr., "Low-Frequency Noise Properties of N-p-n AlGaAs/GaAs Heterojunction Bipolar Transistors," *IEEE Trans. Electron Devices,* Vol. 39, 1992, p. 2383.
[4] Jantsch, O., "A Theory of 1/f Noise at Semiconductor Surfaces," *Solid-St. Electron.,* Vol. 11, 1968, p. 267.
[5] Jaeger, R. C., and A. J. Brodersen, "Low-Frequency Noise Source in Bipolar Junction Transistors," *IEEE Trans. Electron Devices,* Vol. ED-17, 1970, p. 128.
[6] Fonger, W., "A Determination of 1/f Noise Sources in Semiconductor Diodes and Transistors," *Transistor I,* Princeton, NJ: RCA Labs, 1956.
[7] Bell, D. A., *Noise and the Solid State,* New York: Wiley, 1985.
[8] van zer Ziel, A., "Theory of Shot Noise in Junction Diodes and Junction Transistors," *Proc. IRE,* Vol. 43, 1955, p. 1639.
[9] Bell, D. A., *Electrical Noise,* London: Van Nostrand, 1960.
[10] Chen, Y., R. N. Nottenburg, M. R. Panish, R. A. Hamm, and D. A. Humphrey, "Microwave Noise Performance of InP/InGaAs Heterostructure Bipolar Transistors," *IEEE Electron Device Lett.,* Vol. EDL-10, 1989, p. 470.
[11] Schumacher, H., U. Erben, and A. Gruhle, "Noise Characteristics of Si/SiGe Heterojunction Bipolar Transistors at Microwave Frequencies," *Electron. Lett.,* Vol. 28, 1992, p. 1167.
[12] Hawkin, R. J., "Limitations of Nielson's and Related Noise Equations Applied to Microwave Bipolar

Transistors, and a New Expression for the Frequency and Current Dependent Noise Figure," *Solid-St. Electron.,* Vol. 20, 1977, p. 191.

[13] van der Ziel, A., *Solid-State Physical Electronics,* 3rd ed., Englewood Cliffs, NJ: Prentice Hall, 1976, Chapter 15.

[14] Liou, J. J., L. L. Liou, C. I. Huang, and Bayraktaroglu, "A Physics-Based Heterojunction Bipolar Transistor Including Thermal and High-Current Effects," *IEEE Trans. Electron Devices,* Vol. 40, 1993, p. 1570.

[15] Gao, G. B., M.-Z. Wang, X. Gui, and H. Morkoc, "Thermal Design Studies of High-Power Heterojunction Bipolar Transistors," *IEEE Trans. Electron Devices,* Vol. ED-36, 1989, pp. 854–862.

[16] Liou, J. J., *Advanced Semiconductor Device Physics and Modeling,* Norwood: Artech House, 1994, Chapter 3.

[17] Meeks, D., "Fundamentals of Heat Transfer in a Multilayer System," *Microwave J.,* January 1992, pp. 165–172.

[18] Joyce, W. B., "Thermal Resistance of Heat Sink with Temperature-Dependent Conductivity," *Solid-St. Electron.,* Vol. 18, 1975, p. 321.

[19] Bayraktaroglu, B., J. Barrette, L. Kehias, C. Huang, R. Fitch, R. Neidhard, and R. Scheres, "Very High-Power-Density cw Operation of GaAs/AlGaAs Microwave Heterojunction Bipolar Transistors," *IEEE Electron Device Lett.,* Vol. 14, 1993, p. 493.

[20] Yuan, J. S., J. J. Liou, and W. R. Eisenstadt, "A Physics-Based Current-Dependent Base-Resistance Model for Advanced Bipolar Transistors," *IEEE Trans. Electron Devices,* Vol. ED-35, 1988, p. 1055.

# Chapter 7
# Numerical Simulation of HBTs

In the previous chapters, various simple, one-dimensional, and analytic models that describe AlGaAs/GaAs HBT behavior were derived and discussed. To gain more physical insight, this chapter presents numerical studies for the AlGaAs/GaAs single- and multi-finger HBTs. For the single-finger HBT, the effects of graded layer, setback layer, and self-heating on the HBT performance will be investigated. For the multifinger HBT, emphasis will be placed on the impact of different base and collector structures on the HBT characteristics. The analysis is carried out using a two-dimensional device simulator called MEDICI run on a Sun SPARC-2 workstation.

## 7.1 OVERVIEW OF THE MEDICI TWO-DIMENSIONAL DEVICE SIMULATOR

The device simulator MEDICI has been used widely by researchers in universities and industry. Given the proper device structure, doping concentration, and bias condition, MEDICI can solve numerically the electron and hole current equations, the Poisson equation, and the electron and hole continuity equations [1]. It can yield two-dimensional contours and vectors as well as quasi-three-dimensional contours, for mobile carriers, electric fields, potentials, and currents in a semiconductor device. External current-voltage characteristics can also be simulated.

The fundamental semiconductor device equations implemented in MEDICI are [1]

$$\nabla^2 V_i = -(q/\varepsilon)(p - n + N_D^+ - N_A^-) \tag{7.1}$$

$$J_n = q\mu_n \nabla V_i + qD_n \nabla n \qquad J_p = q\mu_p \nabla V_i - qD_p \nabla p \tag{7.2}$$

$$\partial n/\partial t = (1/q)\nabla J_n + G_n - U_n \qquad \partial p/\partial t = -(1/q)\nabla J_p + G_p - U_p \tag{7.3}$$

where the notation have their usual meaning. In addition, carrier transport in high and spatially rapidly varying electric fields is modeled using a self-consistent solution of the above drift-diffusion equations and the following carrier energy balance equations [2]:

$$S_n = -2.5u^{Tn}[J_n/q + \mu_n n \nabla u^{Tn}] \tag{7.4}$$

$$\nabla \cdot S_n = (1/q)J_n \cdot \nabla V_i - 1.5[n(u^{Tn} - u^T)/\tau_e] \tag{7.5}$$

Here $S_n$ is the electron energy flow density; $u^{Tn}$ and $u^T$ represent the electron and lattice thermal voltages, respectively; and $\tau_e$ is the electron relaxation time. Hole energy balance equations have similar expressions. This is called the hydrodynamic model or nonequilibrium model because it allows for the existence of nonequilibrium condition in free carriers.

A comprehensive mobility model that covers both the low-field and high-field regions is used in MEDICI; it consists of the concentration-dependent mobility model for the low-field region and the field-dependent mobility model for the high-field region [1]. The formal model uses the table look-up scheme, which chooses the low-field electron and hole mobilities ($\mu_{n0}$ and $\mu_{p0}$) according to the local impurity concentration $N_{total}(x, y)$, and the latter model calculates the mobilities from the equations

$$\mu_n = \mu_{n0}/[1 + (\mu_{n0}E_n/v_{nsat})^{\beta_n}]^{1/\beta_n} \tag{7.6}$$

$$\mu_p = \mu_{p0}/[1 + (\mu_{p0}E_p/v_{psat})^{\beta_p}]^{1/\beta_p} \tag{7.7}$$

Here $\mu_n$ and $\mu_p$ are the high-field electron and hole mobilities, $v_{nsat}$ and $v_{psat}$ are the saturation drift velocities for electrons and holes ($v_{nsat} = 1.07 \times 10^7$ cm/s and $v_{psat} = 8.34 \times 10^6$ cm/s at room temperature), and $\beta_n = 1.11$ and $\beta_p = 2.64$.

The electron and hole generation and recombination processes are described by the Shockley-Read-Hall (SRH), Auger, and band-to-band radiative statistics.

Two options are also available in MEDICI. One option called the lattice-temperature module allows the effects of lattice heating to be included in the simulation by solving the heat transfer equation in addition to the general equations described above. For example, to compute the spatially dependent lattice temperature $T$, the following heat flow equation is used [1]:

$$\delta c \partial T/\partial t = H + \nabla[\lambda(T)\nabla T] \tag{7.8}$$

where $\delta$ is the mass density, $c$ is the specific heat, $H$ is the heat generation, and $\lambda$ is the thermal conductivity of the material. The heat generation in the semiconductor is modeled using

$$H = J_\text{n} \cdot \xi + J_\text{p} \cdot \xi + H_\text{U} \tag{7.9}$$

where $\xi$ is the electric field and the recombination contribution $H_\text{U}$ is given by

$$H_\text{U} = (U_\text{SRH} + U_\text{Auger} + G^\text{II})E_\text{G} \tag{7.10}$$

$U_\text{SRH}$ and $U_\text{Auger}$ are the SRH and Auger recombination rates, respectively; $G^\text{II}$ is the total generation rate due to impact ionization; and $E_\text{G}$ is the energy bandgap. The electron and hole current equations also need to be modified to account for variation of lattice temperature.

Another option called the heterojunction device module provides the means to simulate semiconductor devices that employ multiple semiconductor materials with varying band structure. The main difference between the homojunction and heterojunction simulations is that the intrinsic Fermi potential in a heterostructure device is in general not a solution to the Poisson equation. This is due to the differences in bandgap $E_\text{G}$, electron affinity $\chi$, and densities of states $N_\text{C}$ and $N_\text{V}$ in adjacent materials. In a semiconductor, the electrostatic potential $V_\text{i}$ is defined by

$$V_\text{i} = \chi + E_\text{G}/2q + (kT/2q)\ln(N_\text{C}/N_\text{V}) \tag{7.11}$$

The intrinsic Fermi potential $\psi$ is then related to $V_\text{i}$ as

$$\psi = \psi_\text{v} + V_\text{i} \tag{7.12}$$

where $\psi_\text{v}$ is the local vacuum potential.

## 7.2 EFFECTS OF GRADED LAYER, SETBACK LAYER, AND SELF-HEATING

The N/p$^+$/n AlGaAs/GaAs single-emitter finger HBT shown in Figure 7.1 is considered, and its geometry and device make-up are given in Table 7.1. Simulations are carried out for the following cases: abrupt junction, graded junction, abrupt junction with a setback layer, and graded junction with a setback layer. Both the graded layer and setback layer have a typical thickness of 150 Å. In addition, results simulated without and with the self-heating effect [3] will be presented and compared.

Figures 7.2(a, b) show the simulated equilibrium energy band diagrams for the abrupt HBT and grade HBT with a setback layer, respectively. It can be seen that the conduction-band discontinuity (spike), which exists in the abrupt junction, has been removed in the graded/setback junction. The built-in potential for the HBT is about 1.5V.

Figures 7.3(a, b) illustrate the simulated lattice temperature contours in the abrupt HBT and graded/setback HBT, respectively, biased at $V_\text{BE} = 1.6$V and $V_\text{CE} = 3.0$V and

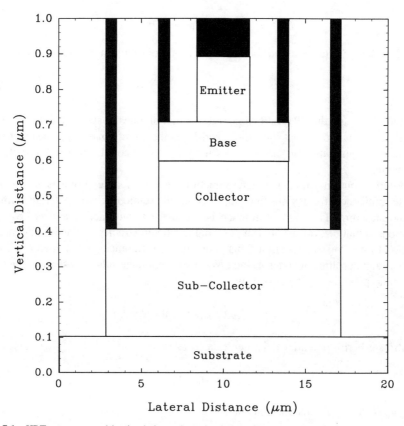

**Figure 7.1** HBT structure used in simulation, where the dark regions represent metal contacts.

with the self-heating effect. The lattice temperature is nearly uniform in the intrinsic HBT region directly underneath the emitter contact (8 μm < $x$ < 12 μm and 0 < $y$ < 0.6 μm) and decreases rapidly toward the ambient temperature (300K) in the extrinsic region and the semi-insulating substrate. Furthermore, it is shown that the peak temperature in the abrupt HBT is smaller than that in the graded/setback HBT. This is because the abrupt HBT has a smaller collector current density than its graded/setback counterpart due to the presence of the spike at the heterointerface (Figure 7.2(a)), as evidenced by the electron current density contours shown in Figures 7.4(a, b) for the abrupt HBT and graded/setback HBT, respectively. In the figures, the two regions where the current density peaks are the collector contacts.

Figures 7.5(a, b) show the lattice temperatures of the abrupt and graded/setback HBTs, respectively, operating at the same collector current density $J_C$ of $8 \times 10^4$ A/cm$^2$. In contrast to the results given in Figures 7.3(a, b) under the same voltage, Figures 7.5(a, b) show that the lattice temperature in the abrupt HBT is slightly higher than that in the

**Table 7.1**
HBT Geometry and Device Make-Up Used in Simulation

|  | Thickness (Å) | Doping Density ($cm^{-3}$) | Type |
|---|---|---|---|
| Emitter | 1000 | $5 \times 10^{17}$ | N-$Al_{0.3}Ga_{0.7}As$ |
| Graded Layer | 0 or 150 | $5 \times 10^{17}$ | Al from 0.3 to 0 |
| Setback Layer | 0 or 150 | intrinsic | p-GaAs |
| Base | 1000 | $1 \times 10^{19}$ | p-GaAs |
| Collector | 3000 | $5 \times 10^{16}$ | n-GaAs |

Emitter Width = 4 µm
Base Width = 12 µm
Collector Width = 12 µm
Semi-Insulating GaAs Thickness = 20 µm

graded/setback HBT (i.e., peak lattice temperatures are 370K and 363K, respectively). This results because of the improved emitter injection efficiency in the graded/setback HBT, which requires a smaller $V_{BE}$ to reach this collector current density than its abrupt counterpart. Consequently, at the same $J_C$ level, the base current density in the graded/setback HBT is smaller than that in the abrupt HBT. Since the power generated in the HBT depends on both the electron and hole current levels (see (7.9)), a lower lattice temperature occurs in the graded/setback HBT, as observed in Figure 7.5.

As mentioned earlier, the hydrodynamic model implemented in MEDICI allows for the existence of a nonequilibrium condition in free carriers, which implies that free carriers can have higher temperatures than that of the lattice if the system is subject to a large perturbation. Figure 7.6 shows the electron temperature contour in the HBT operating at a collector current level of about $10^5$ A/$cm^2$. A peak electron temperature of 1500K is found near the base-collector junction. This high electron temperature, which results from the very large electric field in the base-collector junction, then gives rise to velocity overshoot and enhances the cutoff frequency of the HBT.

At a fixed $V_{BE}$, the HBT base current is also increased by the graded/setback structure, but to a lesser extent than the collector current, as shown in the hole current density contours for the abrupt HBT and graded/setback HBT given in Figures 7.7(a, b), respectively. In the figures, the two regions where the current density peaks are the base contacts. These trends agree with theoretical and experimental studies reported in the literature [4–7]. Note that the self-heating effect is included in the above results.

The current gains for the four different HBTs versus the collector current $I_C$ (in A/mm, where micrometer is HBT's third dimension) simulated without and with self-heating effect are illustrated in Figures 7.8(a, b), respectively. Experimental data measured from the abrupt and graded HBTs are also included in Figure 7.8(b). The devices were fabricated via MOCVD with a wet-etching, self-aligned process. Au, Ti/Pt/Au, Au/Ge/Ni were used for the emitter, base, and collector contacts, respectively. As expected, the self-heating is important in the high current region ($I_C > 10^{-4}$ A/µm) and is negligible

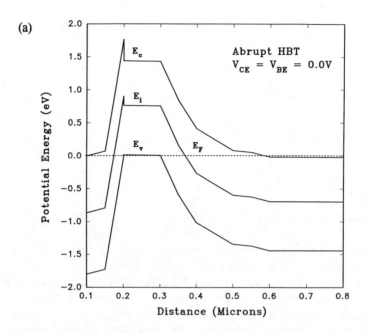

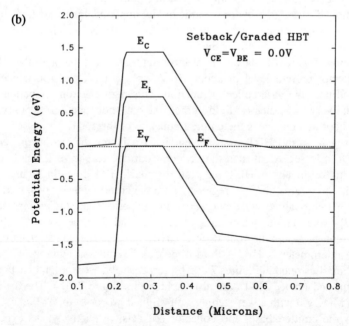

**Figure 7.2** Simulated equilibrium energy band diagrams for (a) an abrupt HBT and (b) an HBT with graded and setback layers.

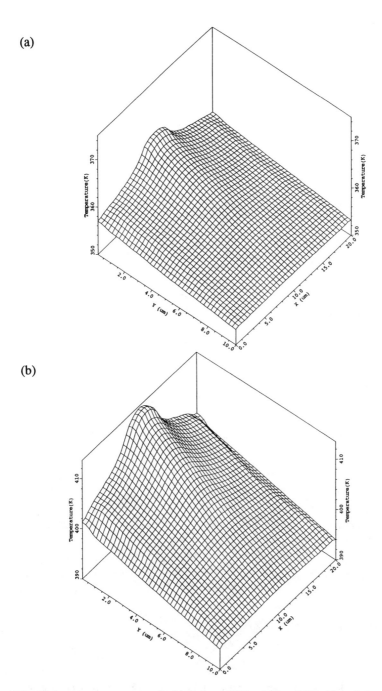

**Figure 7.3** Lattice temperature contours for (a) an abrupt HBT and (b) an HBT with graded and setback layers. The HBTs are biased with $V_{BE} = 1.6V$ and $V_{CE} = 3.0V$.

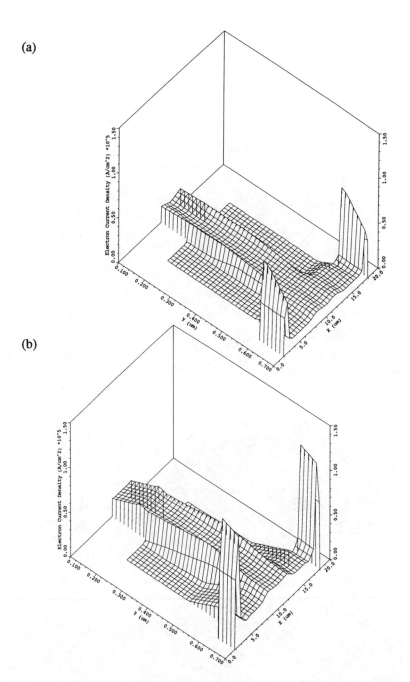

**Figure 7.4** Simulated electron current density contours including self-heating effect for (a) an abrupt HBT and (b) an HBT with graded and setback layers. The HBTs are biased with $V_{BE} = 1.6V$ and $V_{CE} = 3.0V$.

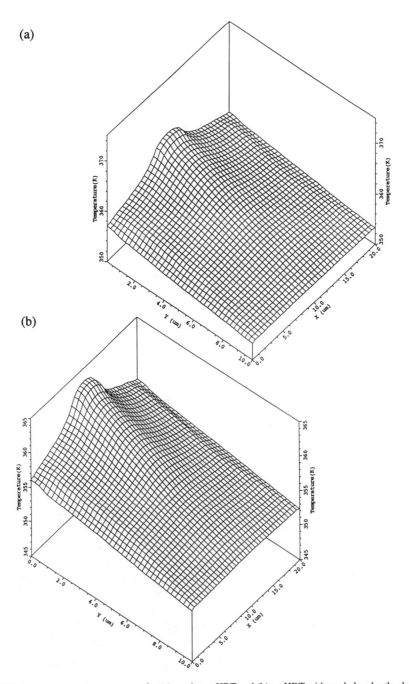

**Figure 7.5** Lattice temperature contours for (a) an abrupt HBT and (b) an HBT with graded and setback layers operating at the same collector current density of $8 \times 10^4$ A/cm$^2$.

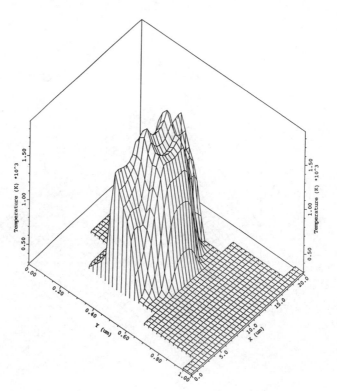

**Figure 7.6** Electron temperature contour in the HBT simulated at a collector current level of $10^5$ A/cm$^2$.

otherwise. Another finding from the results is that while the abrupt HBT has the highest current gain among the four cases when the current is low, it nonetheless has the worst dc performance at the high current level. On the other hand, using a combined graded and setback layer thicknesses of 150 Å yields the highest current gain in the high current region.

The HBT's cutoff frequency at a high current is also degraded significantly by the self-heating, as shown in Figures 7.9(a, b), respectively, because the elevated lattice temperature that results from the self-heating effect makes the Kirk effect more prominent and reduces the free-carrier saturation velocity [3], which subsequently increases the base transit time and the collector signal delay time, respectively. Again, the highest cutoff frequency obtainable from the HBT can be improved considerably by inserting both the graded and setback layers before and after the heterointerface. As shown in Figure 7.9(b), the simulated and measured results are in good agreement.

The following conclusions can be drawn from the above results.

1. The self-heating effect is a key factor limiting the AlGaAs/GaAs HBT current gain and cutoff frequency at high current levels.

(a)

(b)

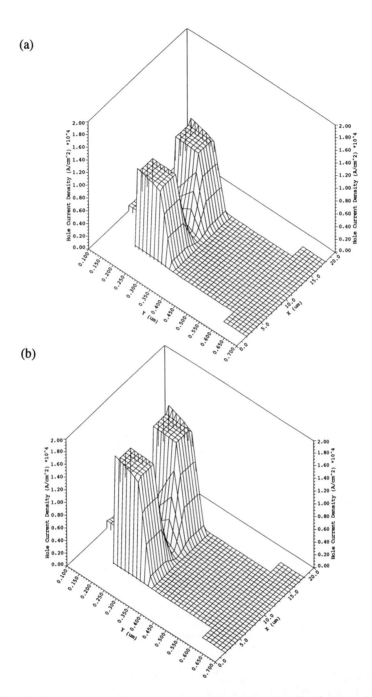

**Figure 7.7** Simulated hole current density contours including self-heating effect for (a) an abrupt HBT and (b) an HBT with graded and setback layers. The HBTs are biased with $V_{BE} = 1.6V$ and $V_{CE} = 3.0V$.

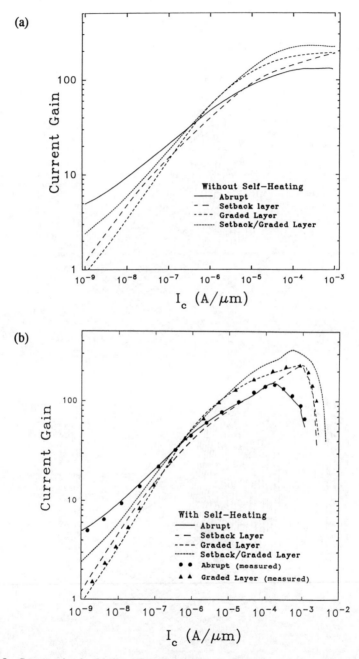

**Figure 7.8** Current gains for the four HBTs simulated (a) without and (b) with self-heating effect. Results measured from the abrupt and graded HBTs are also included.

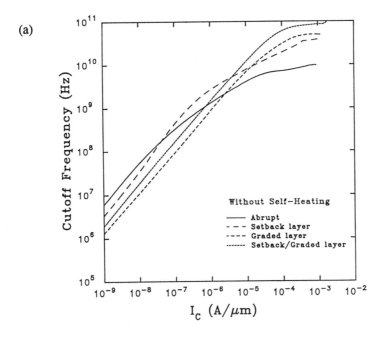

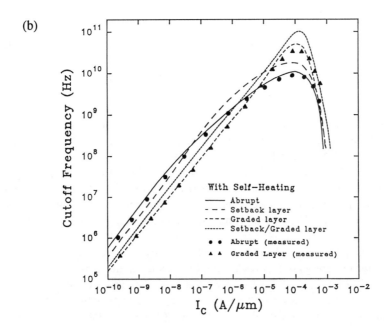

**Figure 7.9** Cutoff frequencies for the four HBTs simulated (a) without and (b) with self-heating effect. Results measured from the abrupt and graded HBTs are also included.

2. The self-heating is more prominent in the graded/setback HBT than abrupt HBT because the abrupt HBT has a smaller emitter injection efficiency and thus a smaller collector current density due to the presence of the conduction band discontinuity at the heterointerface.
3. The elevated lattice temperature in the HBT that results from the self-heating effect decreases the collector current and increases the base current, which then decreases the current gain at the high current region.
4. The inclusion of graded and/or setback layers increases the collector current and, to a lesser extent, increases the base current.
5. Optimal peak current gain and cutoff frequency can be obtained by using both the graded and setback layers.

## 7.3 EFFECTS OF DIFFERENT BASE AND COLLECTOR STRUCTURES

Studies of the multifinger HBT thermal design and analysis have been reported in the literature [8–10]. All of them focus on the multifinger HBT structure in which only the emitter is separated into the island structure (E-island) and both the base and collector are contiguous underneath the emitter fingers, as shown in Figure 7.10(a). In this section, we will also consider two other structures that have emitter/base island (E/B-island) and emitter/base/collector island (E/B/C-island) structures, as shown in Figures 7.10(b) and 7.9(c), respectively. In comparison to the E-island HBT, the E/B-island HBT will have a less uniform lattice temperature and current distribution in the base. Similarly, the E/B/C-island HBT will have less uniform lattice temperature and current distribution in the collector than the other two devices. It is the purpose of the section then to investigate and compare the dc and ac performances of the three multifinger HBTs. As will be shown later, the different island formations considerably affect the dc current gain and cutoff frequency of the HBTs.

Throughout the analysis, all HBT structures considered are N/p$^+$/n Al$_{0.3}$Ga$_{0.7}$As/GaAs/GaAs three-finger HBTs with the following typical make-up: a $5 \times 10^{17}$ cm$^{-3}$ and 1000-Å emitter, 300-Å graded layer, $10^{19}$ cm$^{-3}$ and 1000-Å base, $5 \times 10^{16}$ cm$^{-3}$ and 7000-Å collector, $4 \times 10$ µm$^2$ emitter finger area, and 10-µm finger spacing.

Figures 7.11(a, b) show the current gains β versus the collector current $I_C$ (in A/µm, where micrometers is the width of emitter finger) of the three HBTs simulated at $V_{CE}$ at 2V and 5V, respectively. The current gain at the high current region (i.e., $I_C > 10^{-4}$ A/µm or $10^{-3}$ A) decreases as $V_{CE}$ is increased because a larger $V_{CE}$ gives rise to a more significant thermal effect in the HBT, which subsequently increases the lattice temperature and degrades the current gain. Figures 7.12(a, b) show the lattice temperature contours in the E-island HBT at $V_{CE}$ = 2V and 5V. Not that the intrinsic HBT is located at $0 < y < 0.93$ µm and 234 µm $< x <$ 266 µm, and the lattice temperature decreases rapidly toward the side-walls and semi-insulating substrate. Also, the middle finger is hotter than the outer fingers due to the thermal coupling among the fingers.

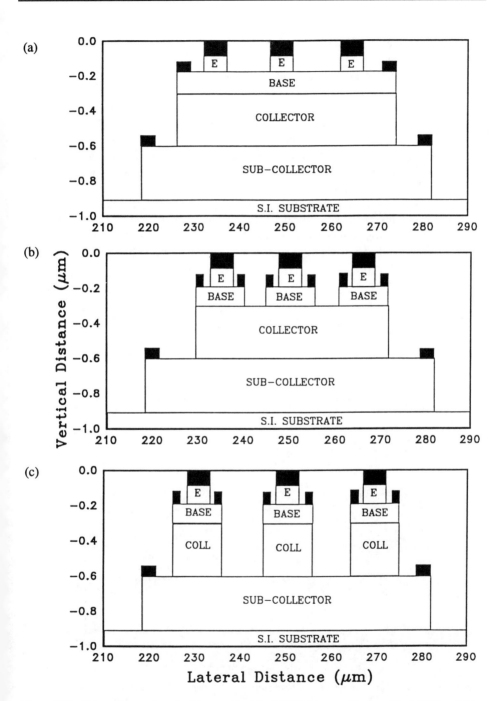

**Figure 7.10** Schematic structures for (a) an emitter-island HBT, (b) an emitter/base-island HBT, and (c) an emitter/base/collector-island HBT.

It can be seen in Figure 7.11(a) that the E-island HBT has the largest $\beta$ for a wide range of current level. At high current levels (i.e., $> 10^{-3}$ A/μm), it is due to the fact that the lattice temperature in such a device is the lowest among the three HBTs, which is evidenced by the lattice temperature contours in Figures 7.13(a–c), which indicate that the E-island and E/B-island HBTs have the lowest and highest lattice temperatures, respectively. At a smaller current (i.e., $10^{-6}$ A/μm), the largest $\beta$ in the E-island HBT results from the contiguous base structure, which makes the electron-hole recombination in the base less prominent and hence reduces the base current. To demonstrate this point, the hole current density contours in the E-island and E/B-island HBTs at a relatively small $V_{BE}$ (i.e., 1.1V) are given in Figures 7.14(a, b), respectively. While the peak hole current density at the base contacts of the E-island HBT appears to be larger than that of the E/B-island HBT, the total base current density of the E/B-island HBT is actually larger because the device has six contacts (i.e., total base current density equals $1.6 \times 10^4$ A/cm$^2$) rather than just two contacts in the E-island HBT (i.e., total base current density equals $1.2 \times 10^4$ A/cm$^2$).

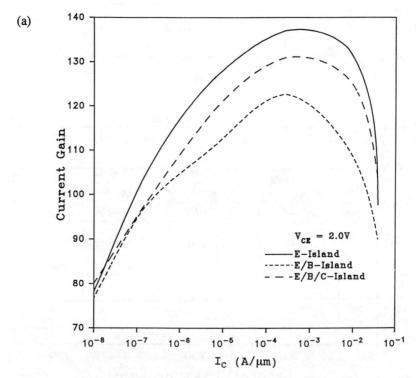

**Figure 7.11** Current gains versus the collector current of the three HBTs simulated at (a) $V_{CE} = 2V$ and (b) $V_{CE} = 5V$.

(b)

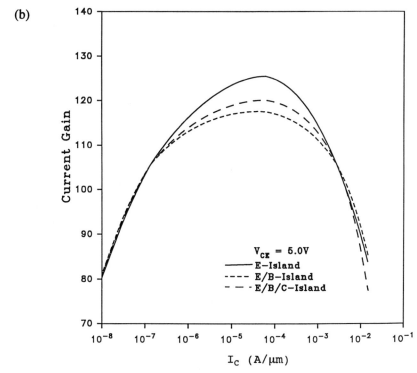

**Figure 7.11** (continued)

Intuitively, one expects that the E/B-island HBT has a lower lattice temperature than the E/B/C-island HBT due to the more uniform collector formation. This is not the case, however, because the contiguous collector in the E/B-island HBT results in a more uniform current contour in the collector and thus a higher collector current. The higher collector current consequently gives rise to a more significant thermal effect and higher lattice temperature in the E/B-island HBT than its E/B/C-island counterpart.

Figures 7.15(a, b) show the cutoff frequencies $f_T$ of the three HBTs simulated at $V_{CE} = 2V$ and 5V, respectively. Like the current gain, the high-current $f_T$ degradation at larger $V_{CE}$ is caused by the thermal effect. At the low current region, however, $f_T$ is slightly higher at larger $V_{CE}$ due to a larger electric field in the base-collector junction and a smaller quasi-neutral base thickness as $V_{CE}$ is increased. Furthermore, the results show that the E/B/C-island HBT has the lowest peak $f_T$, which can be attributed to the fact the E/B/C-island HBT has the least uniform electric field in the base-collector junction among the three devices. This is demonstrated in Figures 7.16(a, b), which plot the electric field

**(a)**

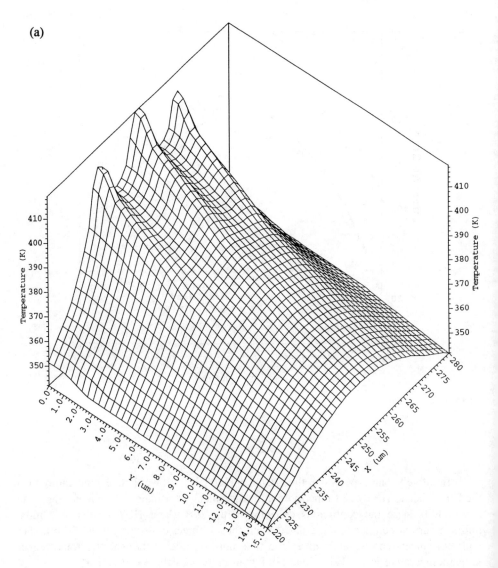

**Figure 7.12** Lattice temperature contours in the E-island HBT at $V_{BE} = 1.5V$ and (a) $V_{CE} = 2V$ and (b) $V_{CE} = 5V$.

(b)

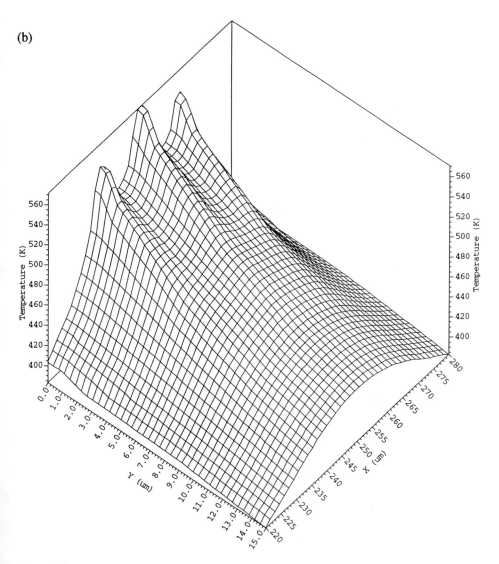

**Figure 7.12** (continued)

**(a)**

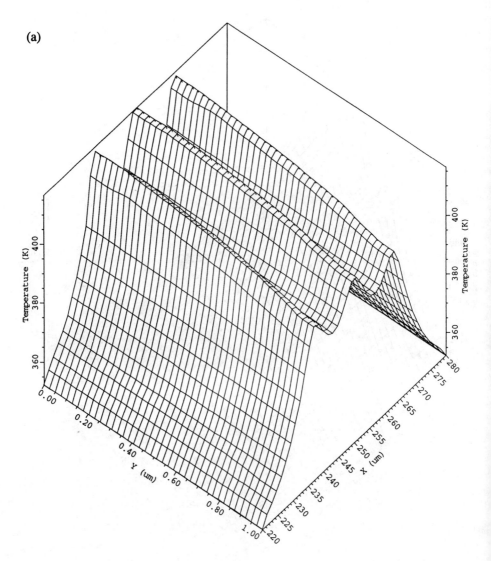

**Figure 7.13** Lattice temperature contours simulated at $V_{BE} = 1.5$V and $V_{CE} = 2$V for (a) an E-island HBT, (b) an E/B-island HBT, and (c) an E/B/C-island HBT.

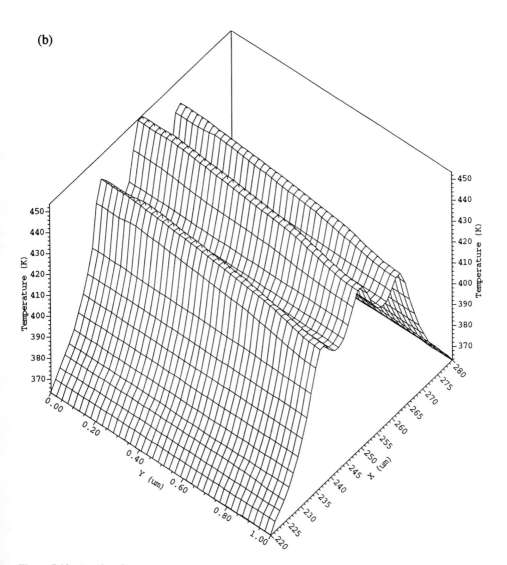

**Figure 7.13** (continued)

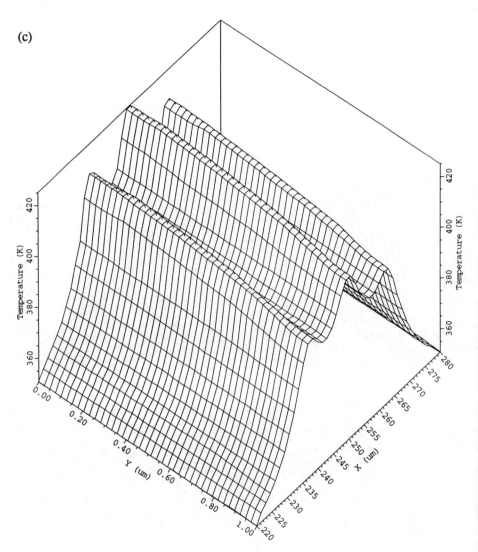

**Figure 7.13** (continued)

(a)

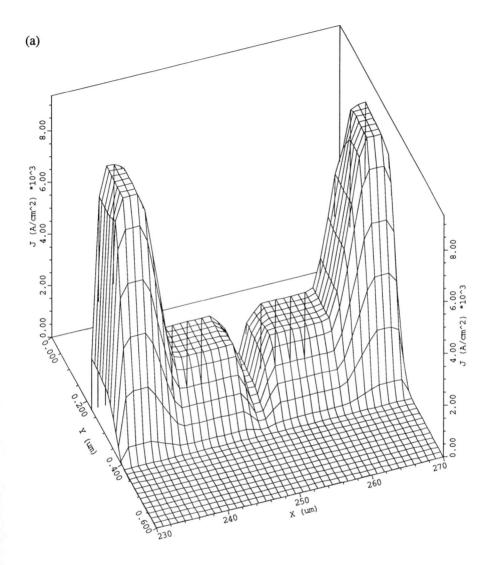

**Figure 7.14** Hole current contours simulated at $V_{BE} = 1.1$ and $V_{CE}$ for (a) an E-island HBT and (b) an E/B-island HBT.

(b)

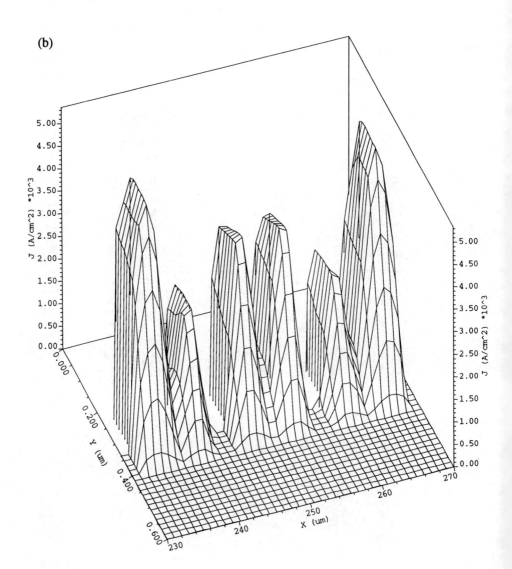

**Figure 7.14** (continued)

contours in the E-island and E/B/C-island and HBTs, respectively. It has been shown in various studies that the free-carrier transit time across the base-collector space-charge region is often the limiting factor for the HBT cutoff frequency [9]. Thus, the lower field in the regions between fingers found in the E/B- and E/B/C-island formations increases the overall base-collector transit time and decreases the overall cutoff frequency of the E/B- and E/B/C-island and HBTs.

Next, we consider the HBT structures in which the base and collector contacts are placed everywhere between the islands, not just placed on the two sides of the base and collector as considered in Figure 7.10. One of the structures (E-island-multicontact) is shown in Figure 7.17(a), and the other structure (E/B/C-island-multicontact) is shown in Figure 7.17(b).

Figures 7.18 and 7.19 compare the $\beta$ and $f_T$, respectively, simulated from the E-island and E/B/C-island HBTs and their two-contact counterparts. The results clearly suggest that placing ohmic contacts between the islands degrades both the high-current dc

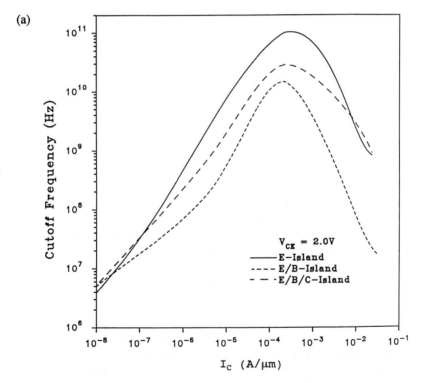

**Figure 7.15** Cutoff frequencies versus the collector current of the three HBTs simulated at (a) $V_{CE}$ = 2V and (b) $V_{CE}$ = 5V.

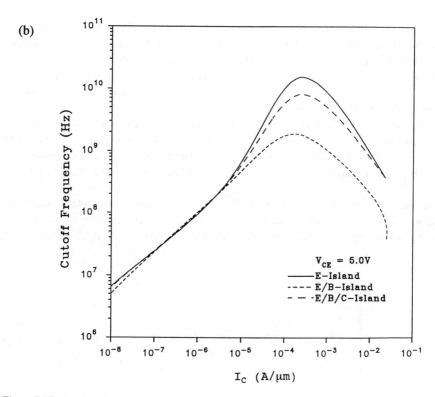

**Figure 7.15** (continued)

and ac performance of the HBTs. This can be attributed to the fact that the voltage drops in the quasi-neutral regions of the multicontact HBTs are smaller compared to those of the two-contact HBTs. As a result, for the same applied voltage, the multicontact HBTs have higher junction voltages than its two-contact counterparts, which then gives rise to a more significant thermal effect and thus degrades the HBT performance. Therefore it is sometimes beneficial to trade the larger voltage drops for a less significant thermal effect in the HBT. The same concept has led to the use of a ballast emitter resistance for HBT thermal design [8].

The following conclusions can be drawn from the study.

1. Among the five structures considered, the E-island HBT is the optimal structure for both high current gain and cutoff frequency applications.
2. For the dc performance, the E/B- and E/B/C-island HBTs are inferior to the E-island HBT because of the higher lattice temperatures in these devices due to the more significant self-heating and thermal-coupling effects.

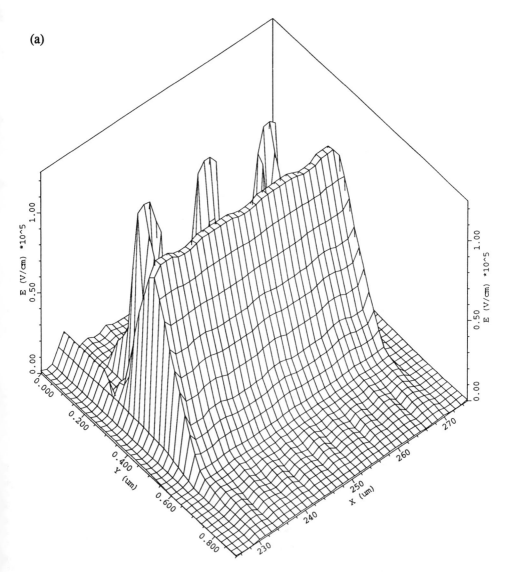

**Figure 7.16** Electric field contours simulated at $V_{CE} = 2V$ for (a) an E-island HBT and (b) an E/B/C-island HBT.

(b)

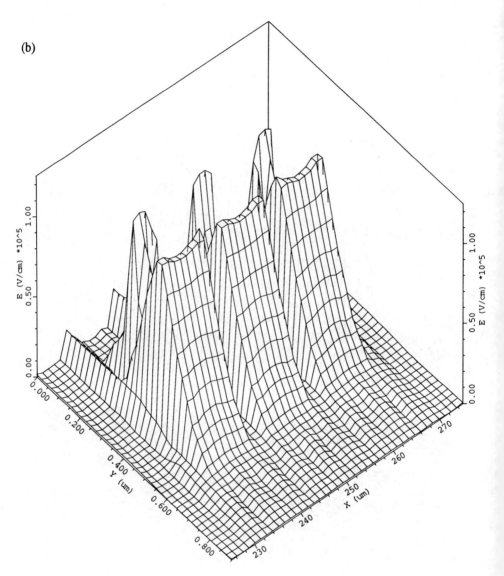

**Figure 7.16** (continued)

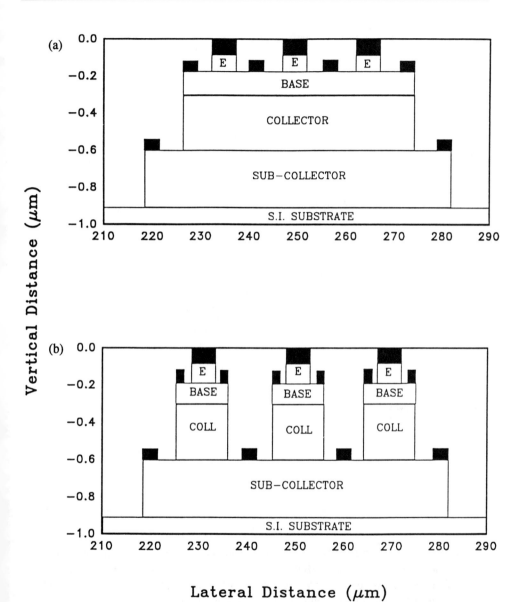

**Figure 7.17** Schematic structures for (a) an E-island-multicontact HBT and (b) an E/B/C-island-multicontact HBT.

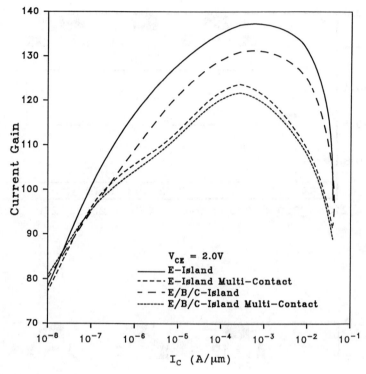

**Figure 7.18** Comparison of the current gains of the E-island, E-island-multicontact, E/B/C-island, and E/B/C-island-multicontact HBTs at $V_{CE} = 2$V.

3. For the ac performance, the E/B- and E/B/C-island HBTs have lower cutoff frequency than the E-island HBT due to the fact that the electric field in the base-collector junction of these devices is much less uniform than that of the E-island counterpart, which then results in a longer overall base-collector transit time and a smaller overall cutoff frequency for the E/B- and E/B/C-island HBTs.
4. Placing ohmic contacts everywhere between the islands (multicontact HBT) degrades the dc and ac performances of the HBT. This is caused by the more significant thermal effect in the multicontact HBT because the voltage drops in the quasi-neutral region of these devices are smaller and the junction voltages are larger compared to those of the two-contact HBT.

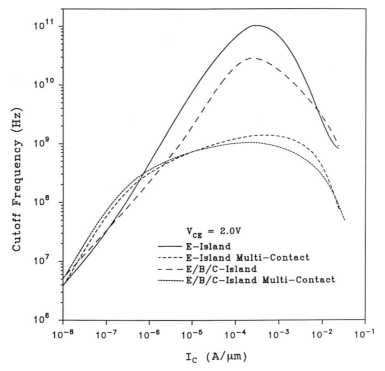

**Figure 7.19** Comparison of the cutoff frequencies of the E-island, E-island-multicontact, E/B/C-island, and E/B/C-island-multicontact HBTs at $V_{CE} = 2$V.

# References

[1] *MEDICI Manual,* Technology Modeling Associates, Inc., Palo Alto, CA, 1993.
[2] Meinerzhagen, B., and W. L. Engl, "The Influence of the Thermal Equilibrium Approximation on the Accuracy of Classical Two-Dimensional Numerical Modeling of Silicon Submicrometer MOS Transistors," *IEEE Trans. Electron Devices,* Vol. 35, 1988, p. 689.
[3] Liou, J. J., L. L. Liou, C. I. Huang, and B. Bayraktaroglu, "A Physics-Based Heterojunction Bipolar Transistor Model Including Thermal and High-Current Effects," *IEEE Trans. Electron Devices,* Vol. 40, 1933, p. 1217.
[4] Das, A., and M. S. Lundstrom, "Numerical Study of Emitter-Base Junction Design for AlGaAs/GaAs Heterojunction Bipolar Transistors," *IEEE Trans. Electron Devices,* Vol. 35, 1988, p. 863.
[5] Ryum, B. R., and I. M. Abdel-Motaleb, "A Gummel-Poon Model for Abrupt and Graded Heterojunction Bipolar Transistors," *Solid-St. Electron.,* Vol. 33, 1990, p. 869.
[6] Chen, S.-C., Y.-K. Su, and C.-Z. Lee, "A Study of Current Transport on p-N Heterojunctions," *Solid-St. Electron.,* Vol. 35, 1982, p. 1311.

[7] Liou, J. J., C. S. Ho, L. L. Liou, and C. I. Huang, "An Analytical Model for the Current Transport in Abrupt HBTs with a Setback Layer," *Solid-St. Electron.*, Vol. 36, 1993, p. 819.

[8] Gao, G. B., M. B. Wang, X. Gui, and H. Morkoc, "Thermal Design Studies of High-Power Heterojunction Bipolar Transistors," *IEEE Trans. Electron Devices*, Vol. ED-36, 1989, p. 854.

[9] Liu, W., S. Nelson, D. G. Hill, and A. Khatibzadeh, "Current Gain Collapse in Microwave Multifinger Heterojunction Bipolar Transistor Operated at Very High Power Densities," *IEEE Trans. Electron Devices*, Vol. 40, 1993, p. 1917.

[10] Kager, A., J. J. Liou, L. L. Liou, and C. I. Huang, "A Semi-Numerical Model for Multi-Emitter Finger AlGaAs/GaAs HBTs," *Solid-St. Electron.*, Vol. 37, 1994, p. 1825.

# Chapter 8
# *Reliability Issues of AlGaAs/GaAs HBTs*

Recently, there has been a great deal of interest in the use of AlGaAs/GaAs HBTs in MMIC power amplifiers, with applications ranging from cellular phones to phased array radars. Keys to the successful use of this device in such applications is a high level of reliability and a good understanding of the failure modes and mechanisms of the HBTs [1]. In this chapter, two topics related to HBT reliability will be discussed. The first is the pre- and post-burn-in base and collector current characteristics of the HBT, and the second topic deals with developing a model to describe the HBT current gain long-term instability.

## 8.1 PRE- AND POST-BURN-IN BASE AND COLLECTOR CURRENTS OF HBTs

A burn-in test carried out in a high temperature/current environment is widely used to assess the reliability of semiconductor devices [2]. Such a test is particularly useful to determine the long-term performance of AlGaAs/GaAs HBTs due to the fact that GaAs properties are susceptible to electrical as well as thermal stresses [3]. Experimental results often show that the burn-in test considerably increases the base current $I_B$ but does not notably alter the collector current $I_C$ [4]. The physics underlying this occurrence has not been fully studied in the past and is not yet well understood.

It has been suggested that defects in a semiconductor can redistribute themselves and diffuse to dislocations due to an enhanced recombination process and/or high thermal energy [5–7]. Energy released by the free carrier during electron-hole recombination at a defect (recombination center) can be localized at the defect. This, together with the ambient thermal energy, may excite the defect to migrate to a defect sink such as dislocation [6]. Since the HBT undergoing the burn-in test is subjected to high-temperature as well as high injection stresses, such a *recombination/thermal enhanced defect diffusion*

(REDF) is a likely kinetics, attributed to the unique post-burn-in $I_B$ behavior. The increase in the number of defects in the base due to the strained lattice is another possible mechanism.

This section provides experimental evidence and physical insights to the pre- and post-burn-in HBT current-voltage characteristics. All current components, including the leakage currents, are investigated; their variations that result from the burn-in test are discussed. The correlation between the experimentally observed HBT behavior and REDF process is then given. An abnormal base current with an ideality factor of about 3 that is observed in the HBT that is subjected to a long burn-in test is also discussed, and a model is developed to describe the physical mechanisms underlying such a current behavior.

## 8.1.1 Theory

Consider a typical mesa-etched, multiemitter finger AlGaAs/GaAs HBT under forward-active operation. To permit a clear focus, we will deal with the free-carrier transport at the emitter-base junction and its periphery, and the effects of the base-collector junction will be neglected. As discussed in Chapter 5, the base and collector currents in the HBT are given by

$$I_B = I_{BN} + I_{BL} \tag{8.1}$$

$$I_C = I_{CN} + I_{CL} \tag{8.2}$$

where $I_{BN}$ and $I_{CN}$ are the normal base and collector currents and $I_{BL}$ and $I_{CL}$ are the base and collector leakage currents generated at the emitter-base perimeter, respectively (see Chapter 5 for the expressions of $I_{BL}$ and $I_{CL}$).

The normal base current $I_{BN}$ consists of (1) the recombination current $I_{SCR}$ in the emitter-base space-charge region, (2) the surface recombination current $I_{RS}$ at the emitter side-walls and extrinsic base surface, (3) the recombination current $I_{RB}$ in the quasi-neutral base region, and (4) the injection current $I_{RE}$ from the base into emitter. Thus

$$I_{BN} = I_{SCR} + I_{RS} + I_{RB} + I_{RE}$$

$$= I_1 \exp(V_{BE}/2V_T) + I_2 \exp(V_{BE}/V_T) + I_n(X_2)(1 - \alpha) + I_p(X_1)\exp(-V_B/V_T) \tag{8.3}$$

where $I_1$ and $I_2$ are the pre-exponential currents for $I_{SCR}$ and $I_{RS}$, respectively, $I_n(X_2)$ is the electron current at the edge of the quasi-neutral base, $I_p(X_1)$ is the hole current at the edge of the quasi-neutral emitter, $V_T = kT/q$ is the thermal voltage, $\alpha$ is the base transport factor, and $V_B$ is the valence-band barrier potential across the emitter-base junction. Note that $I_1$ is determined by the Shockley-Read-Hall recombination process and the trapping density $N_t$ in the space-charge region, $I_2$ is a function of the surface states and location of

Fermi-level pinning, $\alpha$ is proportional to the minority-carrier lifetime $\tau_b$ in the base, and $I_n(X_2)$ and $I_p(X_1)$ depend on the emitter and base doping concentrations, layer thicknesses, and the heterojunction properties. It is important to point out that, unlike all other base current components that are directly proportional to the emitter area, $I_2$ depends on the surface area and thus does not scale in direct proportionality with the emitter area. The normal collector current $I_{CN}$ is given by

$$I_{CN} = I_n(X_2)\alpha \tag{8.4}$$

Note that while electron-hole recombination in the base is accounted for in (8.3) and (8.4) through the base transport factor $\alpha$, $I_{CN}$ and $I_{RB}$ are in fact modeled based on the thin-base case. Such an approach is adequate if the number of defects in the base is relatively low, but it becomes questionable if otherwise.

The base and collector leakage currents that originate at the emitter-base perimeter are affected by the applied base-emitter voltage $V_{BE}$. For discussion, we assume the HBT surface is protected with a nitride layer. As $V_{BE}$ is increased, the potential barriers at the GaAs-nitride and AlGaAs-nitride interfaces are lowered, thus increasing the numbers of electrons in the n-type AlGaAs and holes in the p-type GaAs to surmount the barriers and reach the base and emitter regions, respectively. Once there, these excess minority carriers are no different from those injected across the emitter-base heterojunction and thereby contribute additional base and collector currents to the normal base and collector currents.

The $I_{BL}$ and $I_{CL}$ that originate at the emitter periphery can be expressed analogous to the physical degradation process as [8,9]

$$I_{BL} = P_E J'_{BL}[1 - \exp(-V_{BE}F_L/V_T)] \tag{8.5}$$

$$I_{CL} = P_E J'_{CL}[1 - \exp(-V_{BE}F_L/V_T)] \tag{8.6}$$

where $J'_{BL}$ is the leakage hole current density (A/cm) from the base to emitter, $J'_{CL}$ is the leakage electron current density from the emitter to base, $P_E$ is the emitter perimeter length [$P_E = 2(W_E + L_E)N$, where $W_E$ and $L_E$ are the emitter finger width and length, respectively, and $N$ is the number of emitter fingers], and $F_L$ is an empirical parameter that is related to the activation energy of the degradation process [8]. The values of $J'_{BL}$, $J'_{CL}$, and $F_L$ depend on the device fabrication process used.

### 8.1.2 Illustrations and Discussions

For illustration, we consider a 5-emitter finger AlGaAs/GaAs HBT that has a $5 \times 10^{17}$ cm$^{-3}$ emitter doping concentration, $10^{19}$ cm$^{-3}$ base doping concentration, 1700-Å emitter layer thickness, and 1000-Å base layer thickness. The area of each emitter finger is $2 \times 10$ μm$^2$. Figure 8.1 shows the Gummel plots of the device before burn-in test calculated

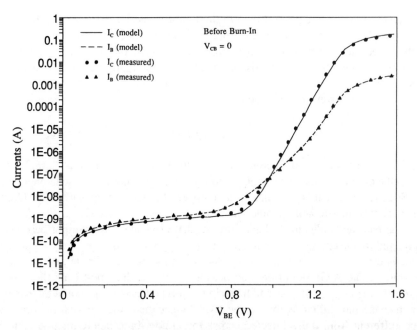

**Figure 8.1** Base and collector currents calculated from the present model and obtained from measurements for a 5-emitter finger (2 × 10 μm² finger area) HBT before the burn-in test.

from the present model and obtained from measurements at base-collector voltage $V_{CB} = 0$. The parameters used in the calculations are listed in Table 8.1. Excellent agreement is found between the model and experimental data. Clearly, at small $V_{BE}$, the base and collector leakage currents are the dominant current components for $I_B$ and $I_C$, respectively, as they increase sharply and become nearly constant as $V_{BE}$ increases (see (8.5) and (8.6)).

The relative importance of the base current components corresponding to $I_B$ in Figure 8.1 is illustrated in Figure 8.2. The results indicate that the base leakage current is the dominant component for $V_{BE} < 0.8V$. Conversely, all $I_{RB}$, $I_{RS}$, and $I_{SCR}$ are important for large $V_{BE}$. At small $V_{BE}$, $I_{RB}$ has the same shape as $I_{BL}$. This is because $I_{RB}$ is proportional to $I_C$ (see (8.3)), which in turn is equal to $I_{CL}$ ($\approx I_{BL}$) under such bias conditions. As $V_{BE}$ increases, $I_{BL}$ becomes negligibly small and $I_{RB}$ returns to its $V_T$-like slope. The surface recombination current $I_{RS}$, like $I_{RE}$, also possesses a $V_T$-like slope and becomes more important as $V_{BE}$ is increased. On the other hand, $I_{SCR}$ has a $2V_T$-like slope due to the dominance of Shockley-Read-Hall recombination process in the space-charge region. These three currents ($I_{RB}$, $I_{SCR}$, and $I_{RS}$) thus give rise to a combined ideality factor $n \approx 1.5$ for the base current at large voltages (Figure 8.1). The current saturation at very large voltages is due to the voltage drops in the series and contact resistances.

Figure 8.3 shows $I_B$ and $I_C$ calculated and measured from the same HBT after a

**Table 8.1**
Parameters Used in Calculating the Pre- and Post-Burn-In HBT Current-Voltage Characteristics

| Parameters | Before Burn-In | After Burn-In (144 hrs) |
|---|---|---|
| $F_L$ | 0.005 | Unchanged |
| $J'_{BL}$ (A/cm) | $1.0 \times 10^{-5}$ | Unchanged |
| $J'_{BL}$ (A/cm) | $0.75 \times 10^{-5}$ | Unchanged |
| $N_t$ (#/cm³) | $5 \times 10^{16}$ | $3 \times 10^{18}$ |
| $\tau_b$ (sec) | $1.0 \times 10^{-10}$ | $1.0 \times 10^{-8}$ |
| $I_2$ (A) | $5.0 \times 10^{-21}$ | $5.0 \times 10^{-23}$ |

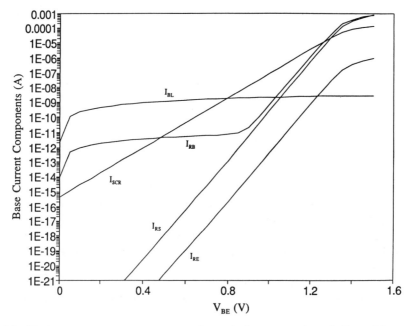

**Figure 8.2** The base current components corresponding to the base current shown in Figure 8.1.

240° C temperature and $10^4$ A/cm² collector current density burn-in test. Experimental data for two different stress periods (144 hrs and 300 hrs) are given. Comparing the results in Figures 8.1 and 8.3, it is apparent that such a test substantially changes $I_B$ but does not notably alter $I_C$. Parameters used in fitting the calculated post-burn-in HBT currents to measurement data are also given in Table 8.1. The two sets of parameters in Table 8.1 indicate that the increased $I_B$ can be attributed to an increase in the trapping density $N_t$ in the space-charge region, which then increases $I_{SCR}$ and gives a relatively large ideality factor for $I_B$. Of equal importance to note is that the pre-exponential current $I_2$ for $I_{RS}$ is

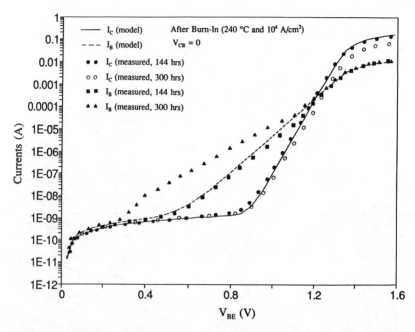

**Figure 8.3** Base and collector currents calculated from the present model (lines) and obtained from measurements (symbols) for the same HBT after the burn-in test.

reduced and the carrier lifetime $\tau_B$ in the base is increased, which subsequently decrease the surface and base bulk recombination currents (see (8.3)). This is due to the fact that both $I_{RS}$ and $I_{RB}$ are proportional to $\exp(V_{BE}/V_T)$, and their importance needs to be minimized to obtain $n \approx 2$ for $V_{BE} > 0.7\text{V}$ observed in the measurements (144-hr stress $I_B$ in Figure 8.3). As a result, it can be suggested that, during the burn-in test, the defects at the surface and bulk of GaAs and AlGaAs redistribute themselves and migrate to dislocations near the emitter-base heterointerface, thus reducing the base and surface recombination currents and increasing the emitter-base space-charge region recombination current.

As illustrated in Figure 8.3, post-stress $I_C$ shows no significant change, except that it decreases slightly in the high bias range as the burn-in period is increased. We believe this is caused by an increase in the post-stress emitter series resistance. Thus, in addition to the good agreement obtained for the 144-hr stress $I_C$, the present model can also closely fit the measured 300-hr stress $I_C$ if a larger emitter resistance is used (not shown in Figure 8.3).

Although the present model predicts quite accurately the 144-hr stress current-voltage behavior and the 300-hr stress $I_C$ characteristics, it nonetheless fails to describe the 300-hr stress $I_B$ between $V_{BE} = 0.3\text{V}$ and $1.1\text{V}$. This base current, which possesses an ideality factor of about 3, cannot be fully explained by any of the base current components discussed so far, which have ideality factors ranging from 1 to 2. According to a recent study [10], this abnormal base current arises from the defects induced by lattice strain (i.e., stress-induced defects) in the base during the long stress test. The results in [10] also show that introducing indium in the carbon-doped base to relax the lattice strain can reduce the post-burn-in base current by a factor of approximately two orders of magnitude.

It is important to point out that the HBT considered does not have an AlGaAs ledge structure. Such a structure passivates the emitter side-walls and extrinsic base surface [11], which then minimizes the migration of surface defects to dislocations due to the thermal/electrical stress. As a result, ledge passivated HBTs may show no appreciable change in their post-stress $I_B$.

### 8.1.3 Base Current of HBT Subjected to Long Burn-In Test

As mentioned above, the model developed in the previous section fails to give an accurate description for the base current after a long burn-in test because, after a long-burn-in test, the number of defects in the QNB is increased significantly due to the strained lattice [10]. As a result, substantial electron-hole recombination occurs in the QNB, and the thin-base assumption employed in the previous section is no longer valid. Thus, for the HBT after a long burn-in test, $I_B = I_{BL} + I_{BN}$, and

$$I_{BN} = I_{SCR} + I_{RS} + I_{RE} + I_{RB} \tag{8.7}$$

where the recombination current in the QNB can be more accurately given by

$$I_{RB} = Aq \int_0^{X_B} U^{SRH}(x)\, dx \tag{8.8}$$

Here $A$ is the emitter area, $x = 0$ and $X_B$ are the boundaries of QNB, and $U^{SRH}$ is the total SRH recombination rate summing the recombination rates at each trapping state $E_{Ti}$ ($i = 1, 2, \ldots, N$, $N$ is the total number of trapping states) and is denoted by

$$U^{SRH} = \sum_{i=1}^{N} U_i^{SRH} \tag{8.9}$$

and

$$U_i^{SRH} = (pn - n_i^2)(N_{Ti}\sigma_i v_{th})/\{p + n + 2n_i \cosh[(E_{Ti} - E_i)/kT]\} \tag{8.10}$$

p and n are hole and electron concentrations in the QNB, $n_i$ is the intrinsic free-carrier concentration, $N_{Ti}$ is the trapping density at $E_{Ti}$, $\sigma_i$ ($\approx 10^{-4}$ cm$^{-2}$) is the capture cross section at $N_{ti}$, $v_{Th}$ ($\approx 10^7$ cm/s) is the electron thermal velocity, and $E_i$ is the intrinsic Fermi energy. The minority-carrier lifetime $\tau_B$ in the p-type QNB is related to the electron concentration as

$$\tau_B = (n - n_0)/U^{SRH} = \Delta n/U^{SRH} \qquad (8.11)$$

where $n_0$ is the equilibrium electron concentration and $\Delta n$ is the excess electron concentration. For a base with an arbitrary length,

$$\Delta n = \Delta n(0)\sinh[(X_B - x)/L_n]/\sinh(X_B/L_n) \qquad (8.12)$$

where $L_n = (D_n\tau_B)^{0.5}$ is the electron diffusion length in the QNB.

Since $\tau_B$ and $\Delta n$ are related to each other, a numerical procedure is needed to calculate $U_{SRH}$, and thus $I_{RB}$, iteratively, provided the parameters associated with the SRH process (i.e., $E_{Ti}$, $N_{Ti}$, and $N$) are specified. As will be shown later, $I_{RB}$ is the current component contributing to the abnormal base current observed in the post-burn-in HBT.

We first investigate the effects of $E_{Ti}$ and $N$ on $I_{RB}$. The device under consideration has a typical make-up of $5 \times 10^{17}$ cm$^{-3}$ emitter doping concentration, 0.15-µm emitter layer thickness, $10^{19}$ cm$^{-3}$ base doping concentration, and 0.1-µm base layer thickness. Also, the intrinsic Fermi energy $E_i$ has been chosen as the reference for $E_{Ti}$ (i.e., $E_{Ti} = 0$ if located at $E_i$). Figure 8.4 shows $I_{RB}$ calculated from the model using fixed

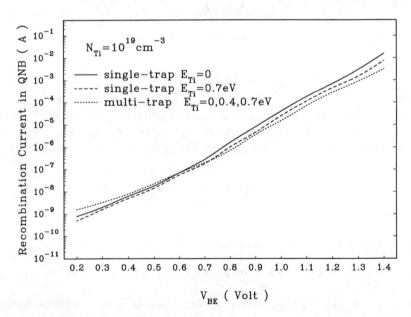

**Figure 8.4** Recombination current in the QNB versus $V_{BE}$ calculated from the model for three different cases of $N_{Ti}$ and $N$.

$N_{Ti} = 10^{19}$ cm$^{-3}$ and a single trap with $E_{Ti} = 0$, a single trap with $E_{Ti} = 0.7$ eV, and multiple traps with $E_{Ti} = 0$, 0.4 eV, and 0.7 eV (i.e., $N = 3$). The results suggest that $I_{RB}$ is insensitive to $E_{Ti}$ and $N$. Furthermore, the current exhibits an $n \approx 3$ characteristic.

Intuitively, one expects that $I_{RB}$ increases with decreasing $E_{Ti}$ and increasing $N$ because $U^{SRH}$ is inversely and directly proportional cosh($E_{Ti}/kT$) and $N$ (see (8.9) and (8.10)), respectively. Nonetheless, $U^{SRH}$ is also a function of the electron concentration in the QNB. Thus, while an decrease in $E_{Ti}$ and an increase in $N$ will tend to increase $U^{SRH}$, such a change will also tend to decrease $\tau_B$ and therefore decrease the electron concentration and $U^{SRH}$ in the region. This compensating mechanism leads to negligible effects of $E_{Ti}$ and $N$ on $I_{RB}$, as observed in Figure 8.4. To demonstrate this, we show in Figure 8.5 the $U^{SRH}$ vs base position calculated for two particular voltages and three different cases of $N_{Ti}$ and $N$. It can be seen that the difference among the areas underneath $U^{SRH}$ for different $N_{Ti}$ and $N$ is about the same for the case of $V_{BE} = 0.6$V and is only slightly larger for the case of $V_{BE} = 1.2$V. This resulted in nearly identical $\int_{QNB} U^{SRH} dx$, and thus $I_{RB}$, for different $N_{Ti}$ and $N$. The dependence of $U^{SRH}$ (x) on $V_{BE}$ is illustrated in Figure 8.6.

Figure 8.7 shows the effect of $N_{Ti}$ on the recombination current in the QNB. Since $I_{RB}$ is insensitive to $E_{Ti}$ and $N$, we have arbitrarily chosen a single trap with $E_{Ti} = 0$ in calculations. Clearly, the value of $N_{Ti;Ti}$ affects $I_{RB}$ significantly, and $N_{Ti}$ will be the main parameter in fitting the model calculations with experimental data.

Figure 8.8 shows the pre- and post-burn-in base currents of HBT-A (device make-up is given in Table 8.2) calculated from the model and obtained from measurements. The

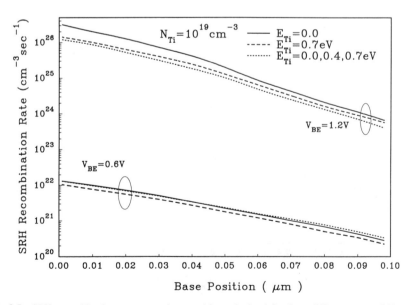

**Figure 8.5** SRH recombination rate versus base position calculated for three different cases of $N_{Ti}$ and $N$.

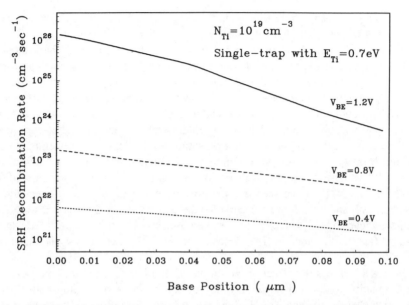

**Figure 8.6** SRH recombination rate versus base position calculated for three different levels of $V_{BE}$.

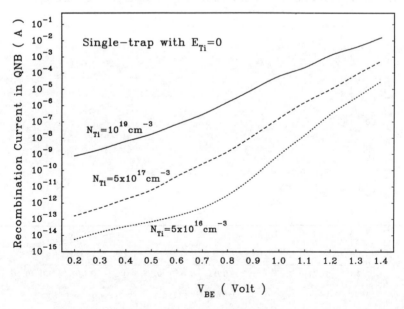

**Figure 8.7** Recombination current in the QNB versus $V_{BE}$ calculated from the model for three different $N_{Ti}$.

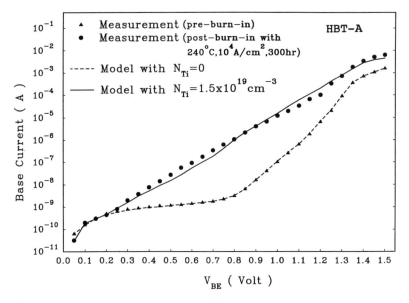

**Figure 8.8** Pre- and post-burn-in base currents of HBT-A calculated from the model and obtained from measurements.

Table 8.2
HBT-A and HBT-B Structures

| Parameters | HBT-A | HBT-B |
|---|---|---|
| Emitter doping ($cm^{-3}$) | $5 \times 10^{17}$ | $5 \times 10^{17}$ |
| Emitter thickness (μm) | 0.17 | 0.18 |
| Emitter area (μm$^2$) | 100 | 30 |
| Base doping ($cm^{-3}$) | $1 \times 10^{19}$ | $1 \times 10^{19}$ |
| Base thickness (μm) | 0.1 | 0.14 |

plateau-like current for $V_{BE} < 0.8V$ in the pre-burn-in HBT is the base leakage current. For the post-stress HBT, the current behavior for $V_{BE} > 0.2V$ is changed to that with $n \approx 3$. This is due to the fact that, in addition to the base leakage current, there is a large $I_{RB}$ in the post-burn-in HBT. $N_{Ti} = 1.5 \times 10^{19}$ cm$^{-3}$ has been used to fit the model to measured data, suggesting the stress-induced defect density in such an HBT is $1.5 \times ^{19}$ cm$^{-3}$. A single trap with $E_{Ti} = 0$ has also been used; like $I_{RB}$, the $I_B$ characteristic is affected only slightly by $N_{Ti}$ and $N$.

Figure 8.9 shows the pre- and post-burn-in base currents of HBT-B (see Table 8.2) calculated from the model and obtained from measurements [10]. For this device, we found that the burn-in test resulted in $N_{Ti} = 7.5 \times 10^{18}$ cm$^{-3}$ in the QNB. This is smaller

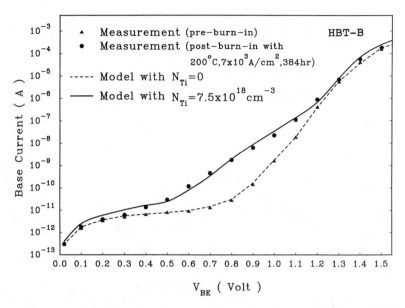

**Figure 8.9** Pre- and post-burn-in base currents of HBT-B calculated from the model and obtained from measurements (*Source*: [10]. © 1995 IEEE.).

than $N_{Ti}$ in HBT-A, due perhaps to the fact that HBT-B is subjected to a less severe burn-in test (200° C and $7 \times 10^3$ A/cm$^2$) than HBT-A (240° C and $10^4$ A/cm$^2$).

## 8.2 MODELING THE HBT CURRENT GAIN LONG-TERM INSTABILITY

Current gain long-term instability in AlGaAs/GaAs HBTs is a severe problem because it sets a limit to the usable lifetime of such devices. Many HBT post-burn-in measurements have shown that while the collector current is affected only slightly by the stress test, the base current can increase considerably due to surface degradation [4,12]. Thus the HBT dc current gain long-term instability is caused mainly by the increase in the recombination current at the extrinsic base surface.

Low-frequency noise has been widely used as a means to evaluate extrinsic base surface integrity [13,14]. For example, most researchers suggested that the 1/*f* noise originates from fluctuations in the occupancy of surface traps, which in turn perturbs the extrinsic base surface recombination current $I_{bs}$ [15,16]. As a result, the magnitude of 1/*f* noise serves as an indicator for the recombination mechanism at the surface. The quality of the base surface, or the interface between dielectric layer and GaAs, on the other hand, is directly related to the base leakage current generated at the emitter-base periphery. Normally, the larger the leakage current, the poorer the surface quality, and the more likely the surface will degrade as time is increased. Such a surface degradation then gives rise to the current gain long-term instability [17].

Previous work has been limited to the determination of the noise sources in the HBT and their origins, and a framework explicitly correlating the initial low-frequency noise and leakage current behavior with the HBT long-term performance is not yet available in the literature.

This section seeks to correlate the $1/f$ noise and base leakage current with the current gain long-term instability in the AlGaAs/GaAs HBT at room temperature. An analytical model is also proposed that, when given the initial noise and leakage current characteristics, can be used to monitor the current gain long-term behavior of HBTs fabricated under the same process condition. Thus, the model is potentially useful for screening unreliable HBTs without having to conduct the long-term stress test.

### 8.2.1 Model Development

A ratio $\Delta h_{FE}/h_{FE0}$ (in %) is often used to characterize the dc current gain long-term instability [17], where $h_{FE0}$ is the initial current gain and $\Delta h_{FE}$ is the accumulated current gain drift over a particular time period $t$. Note that $h_{FE} = I_C/I_B$, where $I_C$ and $I_B$ are the collector and base currents, respectively. The base current of the HBT consists of the intrinsic component $I_{bi}$ and surface component $I_{bs}$:

$$I_B = I_{bi} + I_{bs} \tag{8.13}$$

Since $\Delta h_{FE}$ is caused mainly by the surface degradation, $I_{bi}$ can be assumed constant with respect to time, whereas $I_{bs}$ is a function of time. Thus, for a constant $I_C$ flowing through the device,

$$\Delta h_{FE}/h_{FE0} = \Delta h_{FE}/(I_C/I_{B0}) = \Delta h_{FE}/[I_C/(I_{bi} + I_{bs0})]$$

$$= -(I_{bi} + I_{bs0}) \int_0^t [I_{bi} + I_{bs}(t)]^{-2} \, dI_{bs}(t) \tag{8.14}$$

where $I_{B0} = I_B(t = 0)$ and $I_{bs0} = I_{bs}(t = 0)$.

The remaining task in developing the $\Delta h_{FE}/h_{FE0}$ model in (8.14) is to derive the expressions for $I_{bi}$ and $I_{bs}(t)$.

#### 8.2.1.1 Base Intrinsic Current $I_{bi}$

The base intrinsic current $I_{bi}$ is related to the hole concentration $p(0)$ at the emitter edge of the space-charge region, emitter thickness $W_E$, and base transport factor $\alpha_B$. For a typical HBT with a graded layer that removes the spike at the heterointerface, $p(0)$ can be related to $I_C$, and

$$I_{bi} = (N_B/N_E)(D_p/D_n)(W_B/W_E)I_C\exp(-\Delta E_V/kT) \tag{8.15}$$

where $N_B$ and $N_E$ are the base and emitter doping concentrations, $D_n$ and $D_p$ are the electron and hole diffusion coefficients, $W_B$ is the base thickness, and $\Delta E_v$ is the effective valence-band discontinuity including the graded layer effect.

### 8.2.1.2 Base Surface Current $I_{bs}$

The base surface recombination current $I_{bs}$, on the other hand, depends on the surface property and thus is related to the noise and leakage current. We find that the time-dependent $I_{bs}$ follows the relation

$$I_{bs}(t) = I_{bs0} + I_{bs0}t^\lambda = I_{bs0}(1 + t^\lambda) \tag{8.16}$$

where $\lambda$ is a parameter associated with the base surface quality. The first term on the right-hand side of (8.16) describes the initial condition for $I_{bs}$, and the second term accounts for the time dependence of $I_{bs}$. As will be shown below, $I_{bs0}$ can be determined from the measured initial $1/f$ noise and $\lambda$ is related to the base leakage current.

*Correlation Between $I_{bs0}$ and $1/f$ Noise*

As discussed in Chapter 6, the HBT noise measurements show the existence of three distinct regions in the noise spectra: a $1/f$ shape (Flicker or $1/f$ noise) at lower frequencies, a Lorenztian spectrum (bump or burst noise) at intermediate frequencies, and a constant noise (white or shot noise) at higher frequencies. Our emphasis here will be the $1/f$ noise at low frequencies because it is related to the base surface recombination mechanism and thus the $h_{FE}$ drift.

Assuming the initial surface recombination velocity $S$ ($S \approx 5 \times 10^5$ cm/s without an AlGaAs ledge structure on the base surface and $S \approx 10^4$ cm/s with a ledge) is independent of position and excess free-carrier density at the surface, Fonger [18] showed that the initial base current $1/f$ noise spectral density $S_{Ib}$ can be expressed as

$$S_{Ib}(f) = (I^2_{bs0}/S^2)S_s(f) \tag{8.17}$$

where $S_s(f)$ is the spectral density that characterizes the noise contribution at the surface. Further assume that

$$I_{bs0} \approx qSn(0)L_dP_E \tag{8.18}$$

and that $n(0)$ can be expressed in terms of $I_c$ because

$$I_C \approx qA_E D_n n(0)/W_B \qquad (8.19)$$

Here $L_D$ is the electron lateral diffusion length in the base, $n(0)$ is the injected minority electron carrier at the base edge of the space-charge region, and $P_E$ is the emitter perimeter length. Combining (8.17) to (8.19) yields

$$I_{bs0} = [S_{Ib}(f)fS/C]^{0.5} \qquad (8.20)$$

where $C$ is a constant. It should be pointed out that $S_{Ib}(f)f$ is a constant, and thus an arbitrary $f$ can be selected as long as the $1/f$ noise dominates.

Four AlGaAs/GaAs HBTs (HBT-1, HBT-2, HBT-3, and HBT-4) are studied. HBT-1 and HBT-2 were fabricated in Wright laboratory, Wright-Patterson Air Force Base [19], and both devices have ten 3-μm-diameter emitter dots. They have the same make-up (i.e., mesa-etch structure with an emitter of $5 \times 10^{17}$ cm$^{-3}$ doping density and 1200-Å thickness and a base of $10^{19}$ cm$^{-3}$ doping density and 1000-Å thickness) and dielectric layer (nitride layer) but have different finger patterns (HBT-1 has two 5-dot rows, whereas HBT-2 has one 10-dot row) and were fabricated from different wafers. Their initial noise characteristics have very similar trends as those shown in Chapter 6, only the magnitude and Lorenztian spectrum are somewhat different. At $f = 100$ Hz and collector current density $J_C = 5 \times 10^4$ A/cm$^2$, the base current noise spectral density for the two HBTs is about $8 \times 10^{-19}$ A$^2$/Hz.

The second group of HBTs (HBT-3 and HBT-4) was supplied from an industrial laboratory and was fabricated with a process different from that used in our laboratory. The devices have three $2 \times 20$ μm$^2$ rectangular-shaped emitter fingers. The noise levels are $1.2 \times 10^{-19}$ and $8 \times 10^{-20}$ A$^2$/Hz for HBT-3 and HBT-4, respectively, at $f = 100$ Hz and $J_C = 1.5 \times 10^4$ A/cm$^2$. These noise spectral densities are higher than those in HBT-1 and HBT-2, as they correspond to $1.3 \times 10^{-18}$ and $9 \times 10^{-19}$ A$^2$/Hz at $J_C = 5 \times 10^4$ A/cm$^2$ ($S_{Ib}(f)$ is proportional to $I_C^2$).

*Correlation Between λ and Leakage Current*

Base and collector leakage currents are the dominant current components in AlGaAs/GaAs HBTs operated at relatively small bias voltages. The magnitude of such currents depends strongly on the etching process; quality of the emitter-base and base-collector peripheries, which is covered by a dielectric (e.g., polyimide, and nitride) layer; and quality of the n$^+$-GaAs/semi-insulating-GaAs interface. For discussion, we assume the HBT surface is protected with a polyimide layer. Inferior quality emitter and base peripheries can thus increase the possibility for the free carriers to diffuse to the opposite side of the junction through the GaAs-polyimide and AlGaAs-polyimide interface and subsequently increase the leakage currents.

Since $I_{bs}$ is of interest, we place the emphasis on the base leakage current. An analytical model for the leakage currents in AlGaAs/GaAs HBTs was developed in Chapter 5. The base leakage current $I_{BL,E}$ occurs at the emitter perimeter can be derived analogously to the physical degradation mechanism described in the Arrhenius relationship as

$$I_{BL,E} = P_E J'_{BL,E}[1 - \exp(-V_{BE}F_L/V_T)] \qquad (8.21)$$

where $J'_{BL,E}$ is the leakage hole current density from the base to emitter, $P_E$ is the length of the emitter perimeter, $V_{BE}$ is the applied base-emitter voltage, and $F_L$ is an empirical parameter. The values of $J'_{BL,E}$ and $F_L$ can be obtained by fitting the model to measurement data, and the magnitude of $J'_{BL,E}$ indicates the initial quality of the base surface.

Unlike the leakage process at the emitter-base periphery that requires the free carriers to surmount the potential barrier associated with the dielectric layer, the leakage currents ($I_{BL,SI}$ and $I_{CL,SI}$) at the subcollector/substrate interface are caused by the diffusion of electrons and holes through the leaky interface. As a result, $I_{BL,SI}$ and $I_{CL,SI}$ are independent of the applied voltage.

Figure 8.10 shows the total base currents, including both the leakage and normal components, calculated from the model and obtained from measurements for HBT-1 and HBT-2. The dashed lines in the figure represent the base leakage currents at the emitter-base perimeter. In measurements and calculations, $V_{BC} = 0$ is used to eliminate the leakage component $I_{BL,B}$ at the base-collector perimeter since it is not relevant to the base surface under study. Note that the leakage current dominates the base current for $V_{BE} < 0.8V$. At higher voltages, the base current consists mainly of the normal components such as the hole injection current from the base-to-emitter and recombination current in the base. In the model calculations, $J'_{BL,E} = 5.3 \times 10^{-6}$ A/cm and $3.2 \times 10^{-6}$ A/cm and $I_{BL,SI} = 0$ have been used for HBT-1 and HBT-2, respectively, suggesting that HBT-1 has a poorer base surface initially than HBT-2 and therefore should have a larger surface degradation and a larger current gain drift as time is increased.

The behaviors of HBT-3 and HBT-4 base leakage currents differ considerably from those of HBT-1 and HBT-2, as evidenced by the results shown in Figure 8.11. For small voltages ($V_{BE} < 0.2V$ for HBT-3 and $V_{BE} < 0.4V$ for HBT-4), the base currents of these devices are negative. This is caused by the substantial leakage current ($I_{BL,SI} = 5 \times 10^{-8}$ A for HBT-3 and $I_{BL,SI} = 8 \times 10^{-8}$ A for HBT-4) through the subcollector/substrate interface in these devices. The base leakage current densities at the emitter perimeter are also higher than those of HBT-1 and HBT-2 ($J'_{BL,E} = 5 \times 10^{-5}$ A/cm$^2$ for HBT-3 and $J'_{BL,E} = 4.5 \times 10^{-5}$ A/cm$^2$ for HBT-4), indicating that a large long-term current gain drift is likely in HBT-3 and HBT-4.

We found that $\lambda$ can be correlated with $J'_{BL,E}$ by the empirical relation

$$\lambda = B \cdot (J'_{BL,E})^{0.5} \qquad (8.22)$$

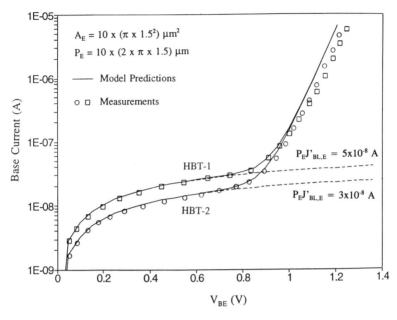

**Figure 8.10** HBT-1 and HBT-2 base currents, including both the leakage and normal currents, calculated from the model and obtained from measurements. The dashed lines represent the base leakage currents at the emitter-base periphery.

where $B$ is an empirical parameter and is related to process signature. For HBTs fabricated with the same process, $B$ should be the same. To determine such a parameter, the value of $\lambda$ is first found by fitting the above model to the long-term current gain measured from a representative device under the stress test. Once the value for $B$ is established, $\lambda$ for any other HBT fabricated from the same process can be found from (8.22) using the same parameter $B$ and initial base leakage current of this particular device. Such a $\lambda$, together with $I_{bs0}$, can then be used to estimate the long-term performance of the HBT. It should be pointed out that when the process is changed, the value of $B$ needs to be redetermined by again measuring the long-term current gain of one HBT fabricated with the new process. The new parameter $B$ is then used in the model to estimate the long-term performance of other HBTs fabricated with this process.

According to the data in Figure 8.10 and the relationship in (8.22), $B \approx 1.5 \times 10^2$ (cm/A)$^{0.5}$ for both HBT-1 and HBT-2, and $\lambda = 0.35$ and $0.27$ for HBT-1 and HBT-2, respectively. For HBT-3 and HBT-4 fabricated from a different process, $B$ needs to be redetermined and is found to be $0.75 \times 10^2$ (cm/A)$^{0.5}$, which leads to $\lambda = 0.53$ for HBT-3 and $\lambda = 0.45$ for HBT-4.

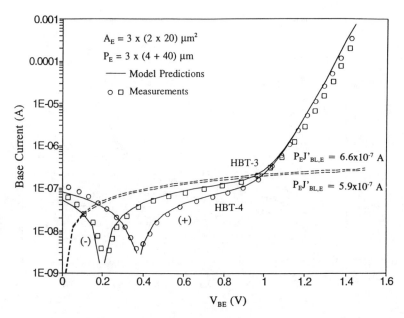

**Figure 8.11** HBT-3 and HBT-4 base currents, including both the leakage and normal currents, calculated from the model and obtained from measurements. The dashed lines represent the base leakage currents at the emitter-base periphery.

### 8.2.2 Results and Discussions

Figure 8.12 shows the current gain variation ($\Delta h_{FE}/h_{FE}$) versus time at room temperature calculated from the model (solid and dashed lines) and obtained from measurements (symbols) for HBT-1 and HBT-2. Two different collector current densities ($2.5 \times 10^4$ and $5 \times 10^4$ A/cm$^2$) are considered. The procedure for the model calculations is as follows. First; the parameter $\lambda$ is determined by fitting the model (solid line) to the measured data of HBT-1 (circles). This, together with $J'_{BL,E}$ of HBT-1 and (8.22), give the value of $B$. The same $B$ and $J'_{BL,E}$ of HBT-2 (Figure 8.10) are then used to find $\lambda$ and the current gain variation of HBT-2 (Figure 8.12). Thus, while the agreement between the model and measurement for HBT-1 is obtained by fitting the data, the agreement found in HBT-2 originates from the physics and correct empirical relations incorporated in the model. Clearly, the current gain drops rapidly at the beginning of the stress test and decreases less quickly as the stress time increases, and the device (HBT-1) that has a higher initial leakage current density at the base surface has a larger current gain drift. Also, it is shown that the gain variation increases with the constant collector current flowing through the device.

Using the same approach discussed in the preceding paragraph, the current gain variations of HBT-3 and HBT-4 at two different stress current levels are calculated and

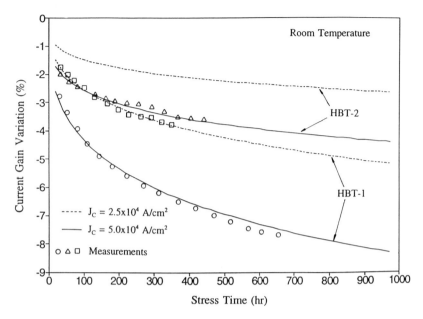

**Figure 8.12** Current gain long-term instability of HBT-1 and HBT-2 calculated from the model (solid and dashed lines) and obtained from measurements (symbols).

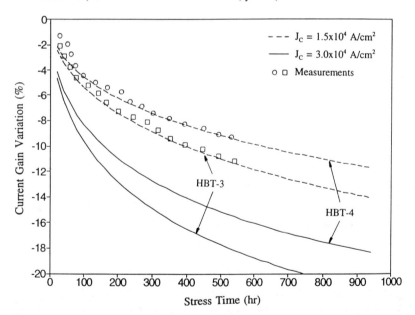

**Figure 8.13** Current gain long-term instability of HBT-3 and HBT-4 calculated from the model (solid and dashed lines) and obtained from measurements (symbols).

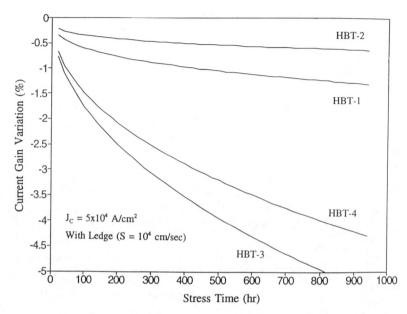

**Figure 8.14** Current gain variation versus time calculated assuming HBT-1, HBT-2, HBT-3, and HBT-4 have an AlGaAs ledge on the base surface.

shown in Figure 8.13 (solid and dashed lines). Again, good agreement is found between the model and measurements. Compared to HBT-1 and HBT-2, these devices have a larger current gain degradation, which is expected because HBT-3 and HBT-4 possess higher base leakage current densities.

All four HBTs considered do not have a passivated base surface. If an AlGaAs ledge structure was incorporated on the base surface of these devices, the initial recombination velocity at the base surface would decrease, which would subsequently reduce the initial base current noise spectra [20] and thus the current gain long-term instability. This is evidenced by the results shown in Figure 8.14, which were calculated from the model assuming HBT-1, HBT-2, HBT-3, and HBT-4 have an AlGaAs ledge on the extrinsic base surface ($S = 10^4$ cm/s has been used). In comparison to the results in Figures 8.12 and 8.13, we found that the ledge has reduced the current gain degradation of HBT-1 and HBT-2 by a factor of 10 and HBT-3 and HBT-4 by a factor of 5.

## References

[1] Henderson, T., D. Hill, W. Liu, D. Costa, H.-F. Chau, T. S. Kim, and A. Khatibzadeh, "Characterization of Bias-Stressed Carbon-Doped GaAs/AlGaAs Power Heterojunction Bipolar Transistors," *IEEE IEDM Digest*, 1994.

[2] Sun, C. J., D. K. Reinhard, T. A. Grotjohn, C. J. Huang, and C. C. W. Yu, "Hot-Electron-Induced

Degradation and Post-Stress Recovery of Bipolar Transistor Gain and Noise Characteristics," *IEEE Trans. Electron Devices,* Vol. 39, 1992, p. 2178.
[3] Leamy, H. J., and L. C. Kimerling, "Electron Beam Induced Annealing of Defects in GaAs," *J. Appl. Phys.,* Vol. 48, 1977, p. 2795.
[4] Hafizi, M. E., L. M. Pawlowicz, L. T. Tran, D. K. Umemoto, D. C. Streit, A. K. Oki, M. E. Kim, and K. H. Yen, "Reliability Analysis of GaAs/AlGaAs HBT's Under Forward Current/Temperature Stress," *Digest IEEE GaAs IC Symp.,* 1990, p. 329.
[5] Benton, J. L., M. Levinson, A. T. Macrander, H. Temkin, and L. C. Kimerling, "Recombination Enhanced Defect Annealing in n-InP," *Appl. Phys. Lett.,* Vol. 45, 1984, p. 566.
[6] Petroff, P. M., and L. C. Kimerling, "Dislocation Climb Model in Compound Semiconductors with Zinc Blende Structure," *Appl. Phys. Lett.,* Vol. 29, 1976, p. 461.
[7] Ballamy, W. C., and L. C. Kimerling, "Premature Failure in Pt-GaAs IMPATTs-Recombination Assisted Diffusion as a Failure Mechanism," *Digest IEEEIEDM,* 1977, p. 90.
[8] Liou, J. J., C. I. Huang, and J. P. Barrette, "Base and Collector Leakage Currents of AlGaAs/GaAs Heterojunction Bipolar Transistors," *J. Appl. Phys.,* Vol. 76, 1994, p. 3187.
[9] Ash, M. S., and H. C. Gorton, "A Practical End-of-Life Model for Semiconductor Devices," *IEEE Trans. Reliability,* Vol. 38, 1989, p. 485.
[10] Sugahara, H., J. Nagano, T. Nittono, and K. Ogawa, "Improved Reliability of AlGaAs/GaAs Heterojunction Bipolar Transistors with a Strain-Relaxed Base," *IEEE GaAs Symp. Digest,* 1995, p. 115.
[11] Liu, W., and J. S. Harris, Jr., "Parasitic Condition Current in the Passivation Ledge of AlGaAs/GaAs Heterojunction Bipolar Transistors," *Solid-St. Electron.,* Vol. 35, 1992, p. 891.
[12] Liou, J. J., and C. I. Huang, "The Base and Collector Currents of Pre- and Post-Burn-in AlGaAs/GaAs Heterojunction Bipolar Transistors," *Solid-St. Electron.,* Vol. 37, 1994, p. 1349.
[13] van der Ziel, A., *Noise in Solid State Devices and Circuits,* New York: Wiley, 1986.
[14] Jantsch, O., "A Theory of $1/f$ Noise at Semiconductor Surfaces," *Solid-St. Electron.,* Vol. 11, 1968, p. 267.
[15] Jaeger, R. C., and A. J. Brodersen, "Low-Frequency Noise Source in Bipolar Junction Transistors," *IEEE Trans. Electron Devices,* Vol. ED-17, 1970, p. 128.
[16] Costa, D., and J. S. Harris, Jr., "Low-Frequency Noise Properties of N-p-n AlGaAs/GaAs Heterojunction Bipolar Transistors," *IEEE Trans. Electron Devices,* Vol. 39, 1992, p. 2383.
[17] Zhuang, Y., and Q. Sun, "Correlation Between $1/f$ Noise and $h_{FE}$ Long-Term Instability in Silicon Bipolar Devices," *IEEE Trans. Electron Devices,* Vol. 38, 1991, p. 2540.
[18] Fonger, W., "A Determination of $1/f$ Noise Sources in Semiconductor Diodes and Transistors," *Transistor* I, Princeton, NJ:RCA Labs, 1956.
[19] Bayraktaroglu, B., R. Fitch, J. Barrette, R. Scherer, L. Kehias, and C. I. Huang, "Design and Fabrication of Thermally-Stable AlGaAs/GaAs Microwave HBTs," *IEEE/Cornell Conf. Proc.,* 1993, pp. 83–92.
[20] Costa, D., and A. Khatibzadeh, "Use of Surface Passivation Ledges and Local Negative Feedback to Reduce Amplitude Modulation Noise in AlGaAs/GaAs Heterojunction Bipolar Transistors," *IEEE Microwave and Guided Wave Lett.,* Vol. 4, 1994, p. 45.

# About the Author

Juin J. Liou received the B.S. (with honors), M.S., and Ph.D. degrees in electrical engineering from the University of Florida, Gainesville, in 1982, 1983, and 1987, respectively.

From 1985 to 1986, he was an instructor in the Department of Electrical Engineering at the University of Florida. In 1987, he joined the Department of Electrical Engineering at the University of Central Florida, Orlando, where he is now an associate professor and graduate coordinator. His current research interests are in CAD tool development, circuit design and simulation, and modeling semiconductor devices, including homojunction bipolar transistors, heterojunction bipolar transistors, metal-oxide-semiconductor field-effect transistors, junction field-effect transistors, solar cells, and photoconductive switching devices.

Dr. Liou has published one textbook, *Advanced Semiconductor Device Physics and Modeling* (with Artech House, Boston) and more than 180 technical papers in refereed journals and international and national conference proceedings and has presented invited seminars or conference papers in several countries. He has held consulting positions with research laboratories and companies in the United States, Germany, Japan, Singapore, and Taiwan. He also serves as a technical reviewer for several journals, serves as a committee chair or member for several international conferences, and is an associate editor for the *Simulation Journal* in the area of VLSI and circuit simulation. He has worked on numerous semiconductor device modeling and integrated circuit simulation projects and has been awarded more than $1.2 million in research grants from industry, state, and federal agencies.

Dr. Liou's honors and awards include Distinguished Research Award, Electrical Engineering Department, University of Central Florida, 1989, 1990, 1992, and 1995; Distinguished Researcher Award, College of Engineering, University of Central Florida, 1992; Senior College Research Award, University of Central Florida, 1993; Faculty Outstanding Award, Student Engineering Council, University of Central Florida, 1993; Engineer of the Year, IEEE Orlando Section, 1992; and Eminent Engineer, Tau Beta Pi, 1992. He was listed in *Who's Who Among Young American Professionals, Who's Who in the South and Southwest, Who's Who in Technology,* and *Who's Who in Science and*

*Engineering.* In the summer of 1992, 1993, and 1994, Dr. Liou was selected for the Summer Research Faculty at Solid State Laboratory, Wright-Patterson Air Force Base, Ohio, where he conducted research on modeling the AlGaAs/GaAs heterojunction bipolar transistor.

Dr. Liou is a senior member of the Institute of Electrical and Electronics Engineers.

# Index

Abrupt HBTs, 17–53
    avalanche-multiplication characteristics of, 37–48
    base current, 21–29
    collector-current, 17–20
    cutoff frequency of, 29–37
    energy band, 4
    free-carrier tunneling, 81
    lattice temperature, 168
    scattering parameters, 48–53
    self heating in, 178
    *See also* HBTs
AlGaAs/GaAs HBTs. *See* HBTs
Avalanche collector current behavior, 37–43
    with base pushout, 40–43
    no base pushout, 38–39
Avalanche multiplication, 37–48
    collector current behavior, 37–43
    for collector doping concentrations, 44
    for collector thicknesses, 45
    current characteristics, 42
    factors, 119
        for I-V curves, 41, 44
        from thermal avalanche model, 121
    importance of, 37
    for junction areas, 45
    reverse base current behavior, 43–48
    in reverse-biased base-collector junction, 110
Ballast resistance, 99, 114, 115
Band discontinuities, 3–6
    conduction, 3
    energy, 5
    valance, 6
Barrier potentials
    for conduction band edge, 86
    for graded junction, 75
    on base side, 77
    of conduction band, 78
    on emitter side, 77
    of valence band, 78
    for setback layer thicknesses, 68
Base contact, 9
Base current
    of abrupt HBTs, 21–29
    after burn-in test, 202
    base emitter voltage vs., 46
    before burn-in test, 200
    calculated from model, 28
    comparisons, 27
    components, 21, 138
    density, 88, 89
        components, 71
        current-voltage characteristics, 156
        model, 69
        noise, 150
    emitter contact resistance, 116
    Flicker noise, 152
    graded/setback structure and, 169
    HBT-1, 213
    HBT-2, 213
    HBT-3, 214
    HBT-4, 214
    large-area HBT, 129
    leakage, 133–148
        at base-collector periphery, 139
        components, 138–148
        effects, 137
        at emitter-base periphery, 139
        HBT-3, 212

Base current (continued)
    HBT-4, 212
    *See also* Leakage current
    multifinger HBTs, 113
    post-burn-in, 197–208
    pre-burn-in, 197–208
    setback layer, 69–72
    shot noise, 152, 153, 154
    space-charge-region recombination current, 25–29
    subjected to long burn-in test, 203–208
    surface recombination current, 22–25
    total, 28, 29
    voltage dependencies, 87
    *See also* Collector current
Base etch, 9
Base grading, 57–64
    current gain effects, 58–62
    early voltage effect, 63–64
    electron density and, 60
    energy band diagram, 58
    *See also* Graded layer; Setback layer
Base intrinsic current, 209–210
Base surface current, 210
Base-collector periphery, 136–137
Base-collector voltage, 65
Base-emitter voltage, 115
    base/collector currents vs., 46
Bipolar transistor (BJT), 63
Burn-in, 197–208
    illustrations and discussion, 199–203
    long, base current subjected to, 203–208
    parameters, 201
    theory, 18–19
Burst noise, 150, 152
    density, 150–151
    temperature dependence, 151
    *See also* Noise

Carrier energy balance equations, 166
Collector charging time, 96
Collector contact, 10
Collector current
    of abrupt HBTs, 17–20
    after burn-in test, 202
    base current, 80–84
    base-emitter voltage vs., 46
    before burn-in test, 200
    characteristics, 20
    collector-emitter voltage vs., 114, 118
    delay time, 154
    density, 87, 88
        lattice temperature and, 101–106
        with/without tunneling, 70, 82
    drift-diffusion model, 17–18
    emitter contact resistance, 116
    graded junction, 79–80
    large-area HBT, 129
    leakage, 133–148
        at base-collector periphery, 139
        components, 138–148
        at emitter-base periphery, 139
        *See also* Leakage current
    multifinger HBTs, 113
    post-burn-in, 197–208
    pre-burn-in, 197–208
    setback layer, 67–68
    shot noise, 152, 153, 154
    thermionic-field-diffusion model, 18–20
    voltage dependencies, 87
    *See also* Base current
Collector etch, 10
Collector-subcollector periphery, 137
Collector-up HBTs, 11–12
Concentration-dependent mobility model, 166
Conduction band
    barrier potential in graded layer, 78
    calculated edges, 79
Constant velocity profile, 34
Current gain
    base grading effects on, 58–62
    cutoff frequency comparison, 91
    E-island structure and, 190
    at high current level, 90
    large-area HBT, 130
    long-term instability, 208–216
    at low current level, 90
    with/without passivation ledge, 94
    self-heating effect and, 174
    simulated as function of passivation ledge, 94
    simulated HBTs, 176
Current-induced base pushout, 30
Current-induced base widening, 40
Cutoff frequency
    collector charging time and, 31
    collector resistance and, 31
    constant velocity profile, 34
    current gain comparison, 91
    current-induced base pushout and, 30
    defined, 29
    delay times, 29–30
    diffusion-dominated base transit time and, 30
    drift velocity and, 33
    E-island structure and, 190
    emitter charge time and, 30
    HBT simulation and, 177

of HBTs, 29–37
from Monte Carlo simulation, 36
piecewise linear velocity profile, 36–37
with/without proton-implanted collector, 97
self-heating effect and, 174
step-like velocity profile, 33, 34–35
Dc
   current gains, 72, 84
   performance, 190
Dielectric layers, 141
Diffusion-dominated base transit time, 30
Drift-diffusion
   equations, 166
   model, 17–18
Early voltage, 63
   base grading effect on, 63–64
   base-collector voltage vs., 65
   illustrated, 63
E/B/C-island, 178
   base-collector junction electric field, 181
   current gain comparisons, 194
   current gain vs. collector current, 180–181
   cutoff frequency comparisons, 195
   cutoff frequency vs. collector current, 189–190
   dc performance, 190
   electric field contours, 191–192
   lattice temperature contours, 184–186
   multicontact, 193
   schematic structure, 179
   *See also* E/B-island; E-island
E/B-island, 178
   current gain vs. collector current, 180–181
   cutoff frequency vs. collector current, 189–190
   dc performance, 190
   hole current contours, 187–188
   hole current density, 180
   lattice temperature, 181
   lattice temperature contours, 184–186
   schematic structure, 179
   *See also* E/B/C-island; E-island
E-island, 178
   current gain, 190
      collector current vs., 180–181
      comparisons, 194
   cutoff frequency, 190
      collector current vs., 189–190
      comparisons, 195
   dc performance, 190
   electric field contours, 191–192
   hole current contours, 187–189
   hole current density, 180
   lattice temperature contours, 182–183, 184–186

   multicontact, 193
   schematic structure, 179
   *See also* E/B/C-island; E/B-island
Electron current density
   base gradings and, 60
   contours, 172
Electron quasi-Fermi level splitting
   applied voltage vs., 8
   emitter-base heterojunction with, 7
Electron temperature contour, 174
Emitter area
   effect on noise behavior, 161–163
   increasing, 162
Emitter charge time, 30
Emitter contact, 9
Emitter contact resistance, 114
   base current, 116
   collector current, 116
Emitter fingers, 118
   5-dot, 1-finger HBT, 157
   5-dot, 2-finger HBT, 157
   5-dot, 5-finger HBT, 157
   pattern and geometry, 139
Emitter side-wall surface recombination
   current, 23–25
Emitter-base periphery, 135–136
Emitter-based heterojunction, 3
   energy band diagram, 1
Emitter-up HBTs, 11–12
Enhanced structures, 57–97
   graded base, 57–64
   graded junction, 72–84
   passivation emitter ledge, 92
   proton-implanted collector, 92–97
   setback layer, 64–72
   setback/graded layers combined, 85–92
   types of, 57
Epitaxial layer growth, 8
Equilibrium free-carrier concentration, 100–101
Extrinsic base surface recombination current, 22–23

Flicker noise, 152
Free-carrier concentration, 100–101
Free-carrier tunneling, 81

Graded layer, 72–84
   "alloy" barrier, 76
   barrier potentials, 75
      on base side, 77
      of conduction band, 78
      on emitter side, 77
      of valence band, 78
   base current, 80–84

Graded layer (continued)
   collector current, 79–80
   collector/base currents compared, 84
   defined, 72
   dielectric permittivity, 73
   electric field, 73
   electron-hole recombination, 80
   energy band diagram, 75
   free-carrier tunneling and, 81, 82
   numerical simulation and, 167–178
   one-dimensional Poisson equation, 73
   schematic illustration, 73
   setback layer combined with, 85–92
   space-charge region, 73
   *See also* Setback layer
Graded/setback HBTs
   lattice temperature in, 168–169
   self-heating in, 178
   *See also* Graded layer; Setback layer

HBTs
   abrupt, 4, 17–53
   avalanche-multiplication characteristics of, 37–48
   collector-up, 11–12
   concept of, 2–3
   current gains, 90
   cutoff frequency of, 29–37
   emitter-up, 11–12
   with enhanced structures, 57–97
   fabrication technology, 7–12
   with graded junction, 72–84
   HBT-1, 139–141, 142, 213, 215
   HBT-2, 141, 143, 213, 215
   HBT-3, 141, 144, 145, 214, 215
   HBT-4, 144, 146, 214, 215
   HBT-5, 144, 147
   HBT-A, 207
   HBT-B, 207, 208
   InP-based, 12–15
   isolation, 10
   large-area, 126–129
   mesa-etch, 109, 133
   microwave, 99
   multifinger, 99, 119, 137
   noise characteristics, 149–163
   non-self-aligned, 8–10
   numerical simulation of, 165–195
   operating between 300K and 500K, 123–130
   output power, 99
   with/without passivation ledge, 93
   planar, 10–11
   reliability issues, 197–216
   schematic, 3
   self-aligned, 10
   with setback layer, 64–72
   Si-based, 12–15
   simulation structure, 168
   single-finger, 119
   six-finger, 121
   small-area, 123–126
   structure, 2–3
   thermal effect, 99–130
   three-finger, 136
Heat flow equation, 166
Heat power, 102, 154–155
   generated, 155, 159
Heterojunction
   device model, 167
   emitter-based, 3
   free-carrier tunneling mechanism, 81
   properties, 3–6
   quasi-Fermi energies in, 6–7
   space-charge region thickness, 5
Heterojunction bipolar transistors. *See* HBTs
Heterojunction field-effect transistors (HFETs), 1
High-frequency noise, 152–163
   emitter area effect, 161–163
   frequency-dependent nature, 152
   minority-carrier delay time and, 162
   model, 152–154
   self-heating and, 162
   thermal effect, 154–161
   *See also* Noise
Hole current contours, 175
   E/B-island, 187–188
   E-island, 187–188
Homojunction bipolar transistors (BJTs), 1
Hydrodynamic model, 169

Impact ionization parameters, 119–120
InP-based HBTs, 12–15
   high-performance potentials, 13
Ion implantation, 11
Isolation, 10
Isothermal model
   calculated results, 104
   falloff prediction, 105
I-V characteristics
   multiplication factors for, 41, 44
   from thermal-avalanche model, 120

Kirchhoff transformation, 102
Kirk effect, 104

Large-area HBTs, 126–130
   base current, 129
   collector current, 129

current gain, 130
Lattice temperature
  abrupt HBTs, 168
  collector current density and, 101–106
  contours, 171, 173
    E/B/C-island, 184–186
    E/B-island, 184–186
    E-island, 182–183, 184–186
  E/B-island, 181
  generated heat power and, 155
  graded/setup HBTs, 168–169
  increase, 104, 118
  plotted, 105
  spatially dependent, 125
  three-dimensional, 126, 127, 128
Leakage currents, 133–148
  at base-collector periphery, 136–137
  at collector-subcollector periphery, 137
  components, 144
  densities, 135
  dielectric layers and, 141
  at emitter-base periphery, 135–136
  HBT-1, 139–141, 142
  HBT-2, 141, 143
  HBT-3, 141, 144, 145
  HBT-4, 144, 146
  HBT-5, 144, 147
  hole density, 136
  importance of, 133
  modeling, 135
  positive portion of, 144
  schematic illustration, 134
  at subcollector-substrate interface, 137–138
  *See also* Base current; Collector current
Long-term instability model, 208–216
  development, 209–214
    base intrinsic current, 209–210
    base surface current, 210
    leakage current correlation, 211–214
    noise correlation, 210–211
  HBT-1, 215
  HBT-2, 215
  HBT-3, 215
  HBT-4, 215
  results, 214–216
  *See also* Current gain
Lorenztian coefficient, 153
Lorenztian spectrum, 151, 161
Low-frequency noise, 149
  extrinsic base surface integrity evaluation, 208
  *See also* Noise

MEDICI simulation, 125, 126, 165–167

carrier energy balance equation, 166
comprehensive mobility model, 166
defined, 165
drift-diffusion equation, 166
heat flow equation, 166
heterojunction device model, 167
hydrodynamic model, 169
semiconductor device equations, 165
*See also* Numerical simulation
Mesa-etch HBTs, 109, 133
Metal-organic chemical vapor deposition (MOCVD), 1
Microwave HBTs, 99
Molecular beam epitaxy (MBE), 1
Monte Carlo simulation, 36
Multifinger HBTs, 99, 119
  base current, 113
  collector current, 113
  current-voltage characteristics, 107
  emitter finger geometry, 137
  heat transfer equations, 108
  schematic, 109
  self-heating effect in, 106–117
  thermal effect in, 99
  thermal-coupling effect in, 106–117
  *See also* HBTs

Noise
  base current density, 150
  burst, 150–151, 152
  characteristics, 149–163
  current spectral density, 160–161
  emitter size and, 159
  Flicker, 152
  high-frequency, 152–163
  low-frequency, 149, 208
  measurement procedure, 155–156
  model, 152–154
  overview, 149–151
  performance index, 154
  shot, 151, 152, 153
  thermal, 152
  thermal effect on, 154–161
  voltage, 160, 162
Noise factor, 154
  minimum, 158, 159, 160, 163
  *See also* Noise
Non-self-aligned HBT, 8–10
  processing sequence, 9
  *See also* HBTs
Numerical simulation, 165–195
  base structure effects, 178–195
  collector structure effects, 178–195

Numerical simulation (continued)
    current gains and, 176
    cutoff frequencies and, 177
    electron current density contours, 172
    electron temperature contour, 174
    equilibrium energy band diagrams, 170
    geometry and device make-up, 169
    graded layer and, 167–178
    hole current density contours, 175
    MEDICI, 165–167
    self-heating and, 167–178
    setback layer and, 167–178
    structure, 168

Overshoot region width, 35

Passivation emitter ledge, 92
    current gain with/without, 94
    HBT with/without, 93
    simulated current gain, 94
Piecewise-linear drift velocity, 32
Piecewise-linear velocity profile, 36–37
Planar HBTs, 10–11
    illustrated, 11
    ion implantation, 11
    *See also* HBTs
Post-burn-in base/collector currents, 197–208
Pre-burn-in base/collector currents, 197–208
Proton-implanted collector, 92–97
    collector charging time, 96
    cutoff frequencies with/without, 97
    HBT structure with, 95

Quasi-Fermi level
    electron splitting, 7, 8
    in heterojunction, 6–7
Quasi-neutral base (QNB), 68
    boundaries, 203
    long burn-in test and, 203
    p-type, 204
    recombination current, 204, 206

Recombination current
    in base, 59
    densities, 81
    density distribution, 23
    electron-hole, 80
    in QNB, 22, 204, 206
    space-charge-region, 25–29
    surface, 22–25
        emitter side-wall, 23–25
        extrinsic base, 22–23
Recombination/thermal enhanced defect diffusion (REDF), 197–198

Reliability issues, 197–216
Reverse base current behavior, 43–48

Scattering parameters
    abrupt HBTs, 48–53
    simulated from SPICE, 49
    of two-port network, 48
Self-aligned HBTs, 10
    illustrated, 10
    *See also* HBTs
Self-heating effects, 99, 155
    abrupt HBT and, 178
    current gain and, 174
    cutoff frequency and, 174
    equilibrium free-carrier concentration, 100–101
    graded/setback HBT and, 178
    in multiemitter finger HBTs, 106–117
    numerical simulation and, 167–178
    in single-emitter finger HBTs, 100–106
    temperature rise, 108
Setback layer, 64–72
    advantages of, 64
    barrier potentials, 68
    base current, 69–72
    base current density, 69
    collector current, 67–68
    defined, 64
    graded layer combined with, 85–92
    numerical simulation and, 167–178
    schematic illustration, 66
    *See also* Graded layer
Shockley-Read-Hall (SRH), 166
    recombination process, 138
    recombination rate, 69
    statistics, 25, 26
Shot noise, 151
    in base current, 152, 153, 154
    in collector current, 152, 153, 154
    *See also* Noise
Si-based HBTs, 12–15
    device structure, 14
    fabrication, 14
    gummel plot, 14
    *See also* HBTs
Signal delay time, 35
Single-finger HBTs, 119
    operating at high temperatures, 123
    *See also* HBTs
Six-finger HBTs, 121
    center temperatures, 122
    *See also* HBTs
Small-area HBTs, 123–126
    gummel plots, 124

with/without self-heating effect, 124
*See also* HBTs
Space-charge layer
 base-collector thickness, 154
 boundaries, 38, 39
 thickness, 49
Space-charge region
 graded junction, 73
 thickness, 66, 67
Space-charge-region recombination current, 25–29
 density vs. base-emitter voltage, 27
 ideality factor, 26
SPICE simulation
 model parameters used in, 53
 with/without parasitic elements, 53
 S-parameter, 49
SRH recombination rate, 205–206
Step-like drift velocity profile, 32
Step-like velocity profile, 34–35
Subcollector-substrate interface, 137–138
Subject finger, 118
 determining, 119, 120
Surface recombination current, 22–25
 emitter side-wall, 23–25
 extrinsic base, 22–23
 *See also* Recombination current

Temperature distribution, 110
 contour plots of, 111–112
Temperature-dependent energy bandgaps, 101
Temperatures
 burst noise dependence, 151
 at center and outer fingers, 117
 at center of 6-finger HBT, 122

emitter-finger, 115
high, operating at, 123–129
Thermal effect, 99–130
 isothermal model, 104
 of multifinger HBT, 99
 numerical model, 104
 on noise behavior, 154–161
 self-heating, 99, 100–106
 thermal-avalanche interacting behavior, 117–122
Thermal noise, 152
Thermal resistance, 110, 155
Thermal runway, 114
 onset of, 115
Thermal-avalanche interacting behavior, 117–122
Thermal-avalanche model
 I-V characteristics of, 120
 multiplication factors, 121
Thermal-coupling effects, 99
 in multiemitter finger HBTs, 106–117
Thermionic-field-diffusion model, 18–20, 67, 79
Thermionic-tunneling-diffusion mechanism, 139
Three-dimensional lattice temperatures, 126
 illustrated, 127, 128
Three-finger HBTs, 136

Valence band
 barrier potential for graded junction, 78
 calculated edges, 79
Velocity profiles, 34–37
 constant, 34
 piecewise-linear, 36–37
 step-like, 33, 34–35